Listed in MAA's
Library Recommendatio
for Undergraduate
Mathematics

AF710103

WITHDRAWN

Differential topology with a view to applications

D R J Chillingworth

University of Southampton

Differential topology with a view to applications

Pitman

LONDON · SAN FRANCISCO · MELBOURNE

PITMAN PUBLISHING LIMITED
39 Parker Street, London WC2B 5PB

FEARON-PITMAN PUBLISHERS INC.
6 Davis Drive, Belmont, California 94002, USA

Associated Companies
Copp Clark Ltd, Toronto
Pitman Publishing New Zealand Ltd, Wellington
Pitman Publishing Pty Ltd, Melbourne

First published 1976
Reprinted 1977
Reprinted 1978

AMS Subject Classifications: (main) 58A05, 34–02
(subsidiary) 26A57, 34C–, 34D30, 58F–, 70–34

Library of Congress Cataloging in Publication Data

Chillingworth, David.
Differential topology with a view to applications.

(Research notes in mathematics; 9)
Includes bibliographical references.
1. Differential topology. I. Title. II. Series.
QA613.6.C48 514′.7 76–28202
ISBN 0-273-00283-X

QA
613.6
C48
1976b

© D R J. Chillingworth 1976

All rights reserved. No part of this publication may be reproduced, stored in a retrieval system, or transmitted in any form or by any means, electronic, mechanical, photocopying, recording and/or otherwise without the prior written permission of the publishers. The paperback edition of this book may not be lent, resold, hired out or otherwise disposed of by way of trade in any form of binding or cover other than that in which it is published, without the prior consent of the publishers.

Reproduced and printed by photolithography
in Great Britain at Biddles of Guildford

Preface

With the increasing use of the language and machinery of differential topology in practical applications, many research workers in the physical, biological and social sciences are eager to learn some of the background to the subject, but are frustrated by the lack of any self-contained treatment that neither is too pure in approach nor is written at too advanced a level. Few people have the time and energy to embark on a systematic study in a field other than their own, and so this book is written with the purpose of making differential topology accessible in one volume as a working tool for applied scientists. It could also serve as a guide for graduate students finding their way around the subject before plunging into a more thorough treatment of one aspect or another. The book should perhaps be read more in the spirit of a novel (with a rather diffuse ending) than as a text-book.

The particular aim is to study the *global qualitative behaviour of dynamical systems*, although there are numerous byways and diversions along the route. A dynamical system is some system (economic, physical, biological ...) which evolves with time. Given a starting point, the system moves within a universe of possible states according to known or hypothesized laws, often describable locally by a formula for the 'infinitesimal' evolution, namely a differential equation. The *global* theory is the theory of all possible evolutions from all possible initial states, together with the way these fit together and relate to each other. *Qualitative* theory is concerned with the existence of constant (equilibrium) behaviour, periodic or recurrent behaviour, and long-term behaviour, together with questions of local and overall stability of the system. Global qualitative techniques, mainly stemming from the work of Henri Poincaré (1854-1912), are important both because precise quantitative theoretical solutions may in general be unobtainable, and because in any case a qualitative model is the basis of a sound mental picture without which mechanical calculation is highly dangerous.

The natural universe of evolution for a dynamical system is often a *differentiable manifold*; the evolution itself is a *flow* on the manifold, and a differential equation for infinitesimal evolution becomes a *vector field* on the manifold. Chapters 1-3 of the book are concerned with defining and explaining these terms, while Chapter 4 goes into the qualitative theory of flows on manifolds, ending with some discussion of bifurcation theory. There is an Appendix on basic terminology and notation for set theory.

Inevitably there are many topics which should have been included or developed but which would have expanded the volume to twice its size. Differential forms are hardly mentioned, singularity theory is only touched upon, and the fascinating terrain of general bifurcation theory for differential equations, including the Centre Manifold Theorem (one of the few really *practical* applications of differential topology), is left largely unexplored. I hope the tantalized reader will be able to follow up these topics via the references given.

Formal prerequisites are kept to a minimum. The ideas from topology and linear algebra that are needed are mostly developed from first principles, so that the basic requirements are hardly more than a familiarity with derivatives and partial derivatives in elementary calculus - although these, too, are defined in the text. The exceptions to this are *complex numbers*, which are assumed to be well-known objects to mathematically-minded scientists, and *determinants* and *eigenvalues* of matrices which may be less well-known to some but are everyday equipment for others. My excuse for this logical inconsistency is lack of space and the need to draw a line somewhere: I felt that it was more important to discuss carefully some of the fundamental ideas about linear spaces upon which the rest of the structure is built than to go on to techniques familiar to many people and in any case quite accessible elsewhere. In the first three Chapters the complex numbers feature mainly in examples and illustrations but in their rôles as eigenvalues they become crucial to the main plot in Chapter 4.

As overall references for the qualitative theory of dynamical systems, I suggest the now historic survey article by Smale [125] and the subsequent very readable lecture notes of Markus [73]. The excellent books on differential equations by Arnol'd [11] and Hirsch and Smale [55] are both directed towards the qualitative theory of flows on manifolds. For background on differential topology a recent and attractive text is Guillemin and Pollack [48]: there is also a forthcoming book by Hirsch* [52]. The fascinating article on applications to fluid mechanics and relativity by Marsden, Ebin, and Fisher [75] is highly recommended (see also the introduction to differential topology by Stamm in the same volume).

The present book grew from a series of lectures given to a mixed audience of pure and applied mathematicians, engineers, physicists and economists at Southampton University in 1973/74. It is through the encouragement of several of these colleagues that I have expanded the lecture notes into book form, and I am grateful to them and others for helpful comments and criticisms. I am particularly indebted to Peter Stefan of the University College of North Wales, Bangor who carefully read the original notes and offered many detailed suggestions for improvement.

* Now appeared: highly recommended

Despite all this assistance, I claim the credit for errors. I would also like to thank Professor Umberto Mosco and Professor Nicolaas Kuiper for hospitality at the Istituto Matematico dell' Università di Roma and the I.H.E.S., Bures-sur-Yvette, respectively, during visits to which I wrote up much of the notes. I am grateful also to Pitman Publishing for their interest in the book and patience during its production, to my wife Ann for tolerating the side-effects, and especially to Cheryl Saint and Jenny Medley for spending many long hours producing such a perfect typescript. Finally, my special thanks go to Les Lander for taking upon himself the task of drawing all the figures in the book, and obtaining such professional results in a short space of time.

David Chillingworth
Southampton, August 1976.

I am grateful to Mike Irwin, Mark Mostow and my father H.R. Chillingworth for kindly pointing out a number of misprints and errors which, fortunately, I have been able to correct for the second printing of the book. Also I would like to take this opportunity to draw attention to the Appendix, which explains some set-theoretic terminology and notation that is used without comment in the text.

D.R.J.C. May 1977.

Contents

1 Basic topological ideas

1.1 THE CONCEPT OF A FUNCTION

We usually visualize a real-valued function of a real variable in terms of its graph:

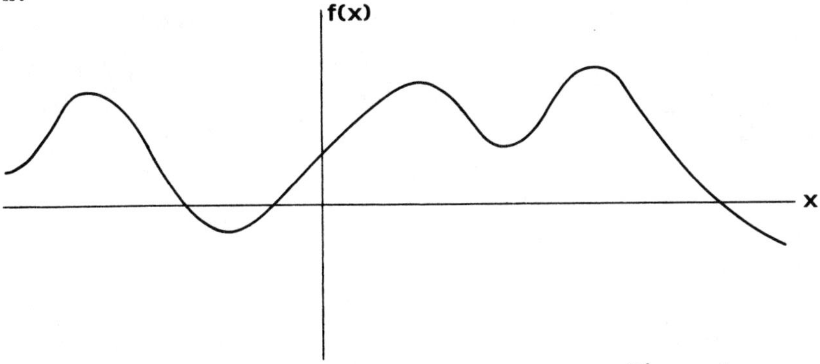

Figure 1

For a given number x (i.e. $x \in R$) the function f provides another real number $f(x)$. The graph consists of all points in a plane coordinatized by (x,y) which satisfy $y = f(x)$, or in formal notation

$$\text{graph}(f) = \{(x,y) \in R \times R \mid y = f(x)\} \quad .$$

Another 'picture' of the function, though less useful than the above, is the following:

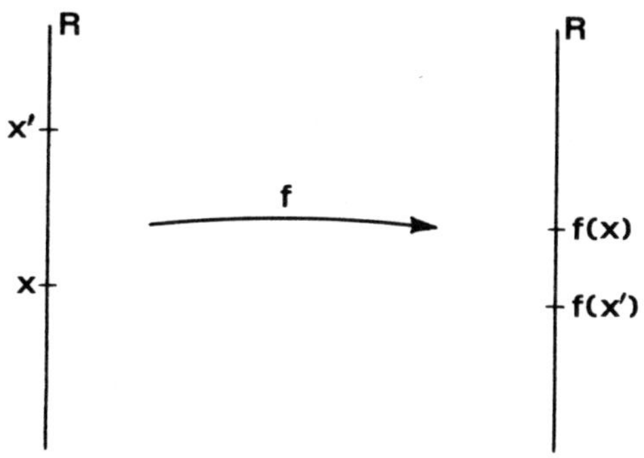

Figure 2

For a real function f of *two* real variables (x_1,x_2) the graph could be thought of as a landscape, with (x_1,x_2) as coordinates in a 'horizontal' plane and $f(x_1,x_2)$ being measured 'vertically'. Formally, we have $(x_1,x_2) \in R \times R = R^2$, $f(x_1,x_2) \in R$ and

$$\text{graph}(f) = \{((x_1,x_2),\ y) \in R^2 \times R = R^3 \mid y = f(x_1,x_2)\} .$$

Alternatively, we can picture f by a 'source and target' picture:

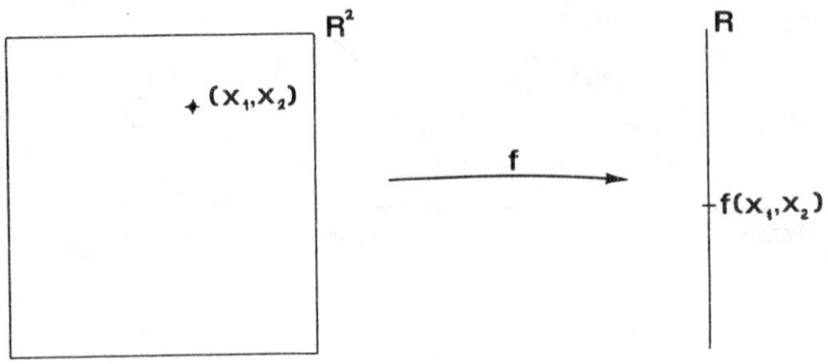

Figure 3

Now suppose we have two functions f_1, f_2 of two variables (x_1,x_2) . We can consider them both at the same time by writing

$$f(x_1,x_2) = (f_1(x_1,x_2),\ f_2(x_1,x_2)) \in R \times R = R^2 ,$$

so that f is a function from R^2 to R^2 .

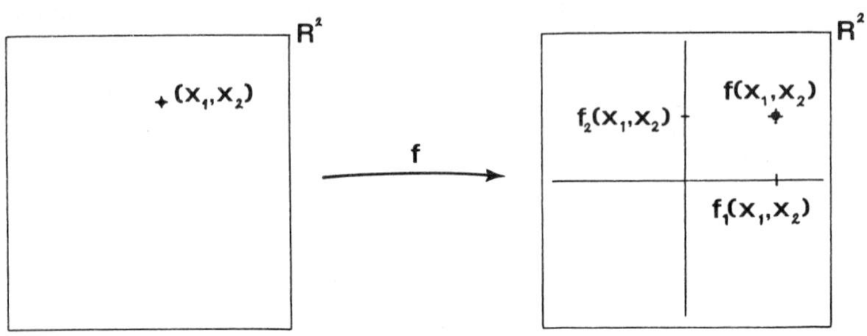

Figure 4

The 'graph' picture in this case is harder to visualize, since by analogy with the previous definition of graph we have

$$\text{graph}(f) = \{((x_1,x_2),\ (y_1,y_2)) \in R^2 \times R^2 \mid y_i = f_i(x_1,x_2)\ ,\ i = 1,\ 2\}$$

and so the graph is a subset of $R^2 \times R^2 = R^4$.

In a similar way, if we are given k functions of n variables we can put them all together to obtain a corresponding function from R^n to R^k . We write $x = (x_1,x_2, \ldots,x_n)$, let

$$f(x) = (f_1(x),\ f_2(x),\ \ldots,\ f_k(x)) \quad ,$$

and keep in mind the picture:

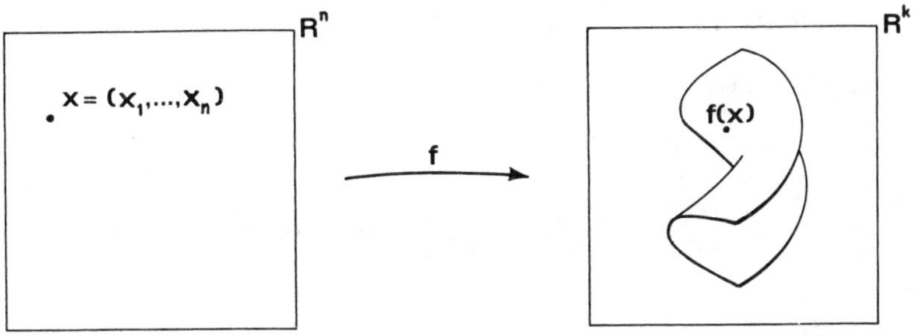

Figure 5

In formal notation this would be written as

$$f : R^n \to R^k \quad .$$

The graph of f would be a subset of $R^n \times R^k = R^{n+k}$, difficult to visualize and in general not yielding as much intuitive information about the behaviour of f as we might extract from Figure 5 by considering the way f causes R^n to be folded up and twisted inside R^k .
The techniques for analyzing such 'folds and twists' are those of topology and calculus together, or *differential topology*. Thus we see already how a study of differential topology may give useful insight into the ways in which k functions of n variables can mutually interact both in general

circumstances and in particular cases. Our aim in the first two chapters will be to develop some of these basic techniques.

Remarks

1. It is frequently necessary to consider functions which are not defined for all values of the variables $(x_1,x_2,\ldots,x_n)$ but only for $x = (x_1,x_2,\ldots,x_n)$ belonging to some subset U, say, of R^n. In this case we of course write

$$f : U \to R^k .$$

2. The word *function* is usually reserved for *real-valued functions* only, i.e. those of the form

$$f : (\text{something}) \to R .$$

In other cases (e.g. for $f : U \to R^k$, $k > 1$) we tend to use the term *map* or *mapping*.

1.2 CONTINUITY

Roughly speaking, a map is continuous if by making a small perturbation in the input (i.e. the independent variables) you obtain only a small change in the output (i.e. the dependent variables). However, this is obviously far too vague a definition. For example, we would wish to think of the function $g : R \to R$ defined by $f(x) = 10^{23}x$ as being continuous (indeed, its graph is a straight line), but it could be argued that small changes in x produce very large changes in $g(x)$. The formal definition of continuity for functions $f : R \to R$ is as follows:

(a) *Continuity at a point* x_o. The function f is *continuous at* x_o if, given any positive number ε (thought of as admissible margin of error in the output), it is then possible to find another positive number δ such that perturbing x_o by less than δ causes $f(x_o)$ to vary by less than ε. Symbolically,

$$|x_o - x| < \delta \text{ implies } |f(x_o) - f(x)| < \varepsilon \quad .$$

(b) *Continuity.* If we are considering f defined on some subset U of R, then we say that f is *continuous on* U if it is continuous at each point x_o belonging to U. If U is the whole of R or is in any case understood from the context then we simply say that f is *continuous*.

Note that the function g in the above example is continuous, since for any x_o and any ε we may take $\delta = 10^{-23}\varepsilon$.

Remarks

1. It is tempting to try to combine (a) and (b) by saying (hopefully) that f is continuous on U if, given any $\varepsilon > 0$, there exists a $\delta > 0$ such that

$$|x - y| < \delta \text{ implies } |f(x) - f(y)| < \varepsilon$$

for all x and y belonging to U. This is *not* the same as the previous definition, however. For example, the function $f : R \to R$ defined by $f(x) = x^2$ is continuous according to (a), (b) but it does not satisfy the 'hopeful' definition. The point is that for the genuine definition of continuity we must allow δ to depend on x_o (as well as on ε, of course). If we do not do this, but demand that the same δ should apply everywhere (given ε), then we have *uniform continuity* - a notion which is in fact of considerable importance in contexts involving approximating functions by other functions as for example in certain techniques of numerical analysis.

2. All 'standard' functions such as polynomials, exponentials, $\sin x$ and so on can easily be proved to be continuous where they are defined. The only hazard to continuity in combining them *ad libitum* is the risk of dividing by a function which vanishes somewhere.

It is easy to see how the definition of continuity will go over to functions $f : R^n \to R^k$, since all that is necessary is to replace the modulus by the euclidean distance from the origin in R^n or R^k (see Example 3 below). However, we shall need to study the 'small change in input gives small change in output' problem in situations where the input and output may be rather more complicated than numbers or n-tuples of real numbers; for example, they may be differential equations, or perhaps collections of functions. Therefore we want to generalize the definition of continuity so that it applies to maps $f : A \to B$ where A and B are sets other than R^n or subsets of R^n. To do this we need notions of *distance* in both sets A, B; then we could simply say

(a) $f : A \to B$ is *continuous at* $x_o \in A$ if, given $\varepsilon > 0$, there exists $\delta > 0$ such that if the distance from x to x_o (in A) is less than δ then the distance from $f(x)$ to $f(x_o)$ (in B) is less than ε;

(b) $f : A \to B$ is *continuous* if it is continuous at every point in A.

Now it turns out that the minimal properties that a 'distance function' d on a set S needs to have in order to reflect adequately the basic relationships of distance in euclidean space are these:-

(1) $d(s,s')$ is always $\geqslant 0$, and $= 0$ when and only when $s = s'$;

(2) $d(s,s') = d(s',s)$ for all s and s' in S ;

(3) $d(s,s'') \leqslant d(s,s') + d(s',s'')$ for all s, s' and s'' in S .

(This last property is known as the *triangle inequality*.) A distance function satisfying (1), (2) and (3) is called a *metric*.

DEFINITION

A function d *which satisfies (1), (2) and (3) above is called a* metric *on* S. *A set* S *together with a particular metric on it is called a* metric space.

Note that d is actually a function from $S \times S$ to R, and not a function on S itself. The image of d is contained in the set R_o^+ of non-negative real numbers (by (1)), and so we could write $d : S \times S \to R_o^+$.

EXAMPLES of metric spaces

1. $S = R$, $d(x,y) = |x - y|$ (the prototype example).

2. $S = R^2$, $d(x,y) = \left|(x_1 - y_1)^2 + (x_2 - y_2)^2\right|^{\frac{1}{2}}$.

3. $S = R^n$, $d(x,y) = \left|\sum_{i=1}^{n} (x_i - y_i)^2\right|^{\frac{1}{2}}$.

Examples 1 and 2 are special cases of Example 3, known as the *euclidean* metric or *usual* metric in R^n .

4. $S = \{\text{bounded functions } f : [a,b] \to R\}$

$$d(f,g) = \sup_{a \leqslant x \leqslant b} |f(x) - g(x)| .$$

5. $S = \{$differentiable functions $f : (a,b) \to R$ with bounded derivative$\}$

$$d(f,g) = \sup_{a < x < b} |f(x) - g(x)| + \sup_{a < x < b} |f'(x) - g'(x)| .$$

Examples 4 and 5 are simple examples of *function spaces*, i.e. spaces whose elements are themselves functions or maps defined on other spaces. Example 6 is also a type of function space.

6. S = {systems of differential equations of the form

$$\begin{aligned} \dot{x}_1 &= f_1(x_1,x_2) \\ & \qquad\qquad (F) \\ \dot{x}_2 &= f_2(x_1,x_2) \end{aligned}$$

defined on R^2 and such that f_1, f_2 are bounded continuous functions}

$$d(F,G) = \sup_{(x_1,x_2)\,\varepsilon\,R^2} \Big| (f_1(x_1,x_2) - g_1(x_1,x_2))^2 + (f_2(x_1,x_2) - g_2(x_1,x_2))^2 \Big|^{\frac{1}{2}}$$

where F, G are defined by f's, g's respectively.

There are two important general types of example:

7. Let S be any set, and let T be a subset of S. Suppose that S is equipped with a metric d. Then T naturally inherits a metric from d, called the *induced* metric. Formally, if we write

$$d : S \times S \to R_o^+$$

then the induced metric is the restriction $d|T \times T$.

8. Let S be any set, and define d by

$$\left.\begin{aligned} d(x,x) &= 0 \quad \text{for all} \quad x \,\varepsilon\, S \\ d(x,y) &= 1 \quad \text{whenever} \quad x \neq y \end{aligned}\right\} .$$

It is easy to verify that this satisfies the rules (1), (2), (3) for a metric. It is known as the *discrete metric* on S. Note that we do not assume S to be a finite set.

We are now in a position to make our formal definition of continuity in the context of metric spaces.

DEFINITION

Let A, B *be metric spaces, with metrics* d_A, d_B *respectively, and let* $f : A \to B$ *be a map.*

(a) f *is* continuous *at* $x_o \in A$ *if, given* $\varepsilon > 0$, *there exists* $\delta > 0$ *such that*

$$d_A(x,x_o) < \delta \quad \textit{implies} \quad d_B(f(x),f(x_o)) < \varepsilon \ .$$

(b) f *is* continuous *if it is continuous at* x_o *for every point* $x_o \in A$.

1.3 CONTINUITY FROM A MORE GENERAL VIEWPOINT

The definition that we have proposed for continuity of maps between metric spaces naturally depends on the choices of the particular metrics. On the other hand the continuity or otherwise of a given map cannot depend too strictly on the precise choice of metric, since it will clearly not matter if we alter the metric in some harmless way such as multiplying all distances by a universal constant factor (corresponding to changing the units of measurement). The closer study of continuity and its independence from specific choices of metric leads naturally to the idea of a *topology*.

We will begin by re-casting the definition of continuity. Given any point x in the metric space S with metric d, let $B_\alpha(x)$ denote the set of all points in S whose distance from x is less than α, i.e.

$$B_\alpha(x) = \{x' \in S \mid d(x,x') < \alpha\} \ .$$

By analogy with 3-dimensional euclidean space we call $B_\alpha(x)$ the *open* α-*ball* around x. Now we have already defined a map $f : A \to B$ between two metric spaces A, B (metrics understood) to be continuous at x_o in A precisely if *given* $\varepsilon > 0$ *there exists* $\delta > 0$ *such that*

$$f(B_\delta(x_o)) \subset B_\varepsilon(y_o) \qquad (*)$$

where y_o *denotes* $f(x_o)$ *in* B. We can rewrite (*) as

$$B_\delta(x_o) \subset f^{-1}(B_\varepsilon(y_o)) ,$$

expressing the fact that not only does x_o lie in $f^{-1}(B_\varepsilon(y_o))$ (obviously, since $f(x_o) = y_o$) but a whole δ-ball around x_o also lies in the subset $f^{-1}(B_\varepsilon(y_o))$ of A. See Figure 6.

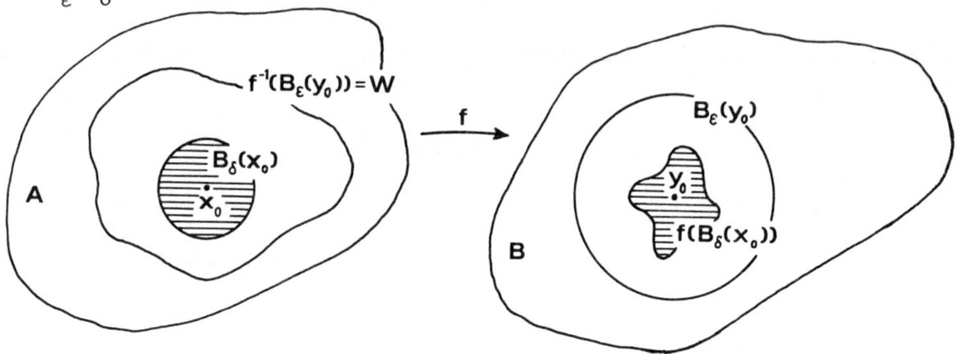

Figure 6

Let us look further at this set $f^{-1}(B_\varepsilon(y_o))$, which we will denote by W. Take another point x_1 in W apart from x_o. Suppose we are given that f is also continuous at x_1. Then it is possible to find some δ'-ball around x_1 lying entirely in W. The proof of this is very simple: all we have to do is to show that $f(B_{\delta'}(x_1)) \subset B_\varepsilon(y_o)$ for some $\delta' > 0$. Let $f(x_1) = y_1$. Then $y_1 \in B_\varepsilon(y_o)$ since $x_1 \in W$, and thus $d(y_1, y_o) < \varepsilon$. Take $\varepsilon' = \varepsilon - d(y_1, y_o)$ and choose δ' such that $f(B_{\delta'}(x_1)) \subset B_{\varepsilon'}(y_1)$: this δ' exists by the continuity of f at x_1. Then the triangle inequality shows immediately that $B_{\varepsilon'}(y_1) \subset B_\varepsilon(y_o)$, which ends the proof. Therefore, if f is continuous everywhere we see that W has the following property: *given any point* x *in* W *there exists some* $\delta > 0$ *such that the ball* $B_\delta(x)$ *is contained in* W. This leads to a general definition:

DEFINITION

Let S *be a metric space. A subset* W *of* S *is an* open set *if, given any* x *in* W, *there exists some* $\delta > 0$ *with* $B_\delta(x)$ *contained in* W.

For example, the interval $\{x \mid 2 < x < 3\}$ is an open set in the real line $\mathbb{R}$, whereas the interval $\{x \mid 2 < x \leq 3\}$ is not (because $x = 3$ fails to lie in any δ-ball contained in the interval). Observe that the line segment $\{(x,y) \mid 2 < x < 3, y = 0\}$ is *not* an open set in $\mathbb{R}^2$, since every δ-ball will now have to contain points with $y \neq 0$. On the other hand

$$\{(x,y) \mid 2 < x < 3, -1 < y < 1\},$$
$$\{(x,y) \mid x^2 + y^2 \neq 1\}$$

are both examples of open sets in $\mathbb{R}^2$. In general a set described by a finite number of strict inequalities will tend to be an open set in $\mathbb{R}^n$ - for reasons which will become very clear shortly.

With this new terminology available we can now re-express the definition of continuity for maps between metric spaces in an extremely simple way.

THEOREM

A map $f : A \to B$ *between metric spaces is continuous if and only if* $f^{-1}(V)$ *is an open set in* A *whenever* V *is an open set in* B.

Important note

The theorem does *not* say that if U is open in A then $f(U)$ is open in B. Consider for example $f : \mathbb{R} \to \mathbb{R}$ given by $f(x) = x^2$, with $U = \mathbb{R}$: then $f(U) = \mathbb{R}_o^+$ which is not open in $\mathbb{R}$.

Proof of the theorem

Suppose first that we are given that f is continuous. Let V be an open set in B : we must show that $f^{-1}(V)$ is open in A. Choose any x in $f^{-1}(V)$. Then $f(x) \in V$, and so since V is open we can find some $\varepsilon > 0$ with $B_\varepsilon(f(x)) \subset V$. Now the continuity of f guarantees the existence of some $\delta > 0$ with $f(B_\delta(x)) \subset B_\varepsilon(f(x)) \subset V$, so $B_\delta(x) \subset f^{-1}(V)$. This argument applies to each $x \in f^{-1}(V)$, so we have shown that $f^{-1}(V)$ is open.

Conversely, let us assume the property about open sets and deduce the continuity of f. Choose a point $x \in A$. For any $\varepsilon > 0$ the ε-ball $B_\varepsilon(f(x))$ is an open set in B (an easy exercise, again using the triangle inequality), and so by hypothesis $f^{-1}(B_\varepsilon(f(x)))$ is open in A. This means that since $x \in f^{-1}(B_\varepsilon(f(x)))$ there is some $\delta > 0$ with $B_\delta(x) \subset f^{-1}(B_\varepsilon(f(x)))$ or, in other words, $f(B_\delta(x)) \subset B_\varepsilon(f(x))$. This applies for each $\varepsilon > 0$, and so proves the continuity of f at x. Since x was an arbitrary point of A we have shown that f is continuous.

In view of this theorem it is clear that in the study of continuity of maps between metric spaces it is the *family of open sets* in each space which is important, rather than the actual metric. More precisely, if two different metrics give rise to the same family of open sets then any map which is continuous using one metric will automatically be continuous using the other.

The family of open sets of a metric space is called its *topology*. All that is needed, therefore, in order to define the concept of continuity of maps from any set A to another set B (whether or not they are given as metric spaces) is the notion of a topology for each of A and B.

Naïvely, we could select any family F of subsets of A and postulate that F shall be the topology for A. Not surprisingly this does not lead very far, unless we insist that the family F obey a few simple rules which the family of open sets in a metric space does obey. The rules which turn out to be crucial are these:

(1) If U and V belong to F then so does $U \cap V$.

(2) If each U_λ belongs to F, for some family $\{U_\lambda\}_{\lambda \in \Lambda}$, then so does $\bigcup_{\lambda \in \Lambda} U_\lambda$.

For convenience we also add a formal clause:

(3) Both the whole set A and the empty set $\emptyset$ belong to F.

Note that clause (1) implies that the intersection of any *finite* number of sets in F also belongs to F, but it does not guarantee anything about the intersection of an infinite number of them. On the other hand, clause (2) asserts that the union of an arbitrary number of sets in F must still belong to F.

It is straightforward to verify (making good use of the triangle inequality) that the family of *open sets* in a metric space does satisfy (1), (2) and (formally) (3). This means that we are now able to define what we mean by a topology on an arbitrary set, whether or not we are given a metric.

DEFINITION

Let S *be a set. A family* F *of subsets satisfying the rules (1), (2) and (3) above is called a* topology *for* S. *The set* S *together with its topology is a* topological space.

By analogy with the metric space situation, the sets in the topology of S are referred to as *open sets*.

EXAMPLES of topological spaces

1. Obviously, every metric space is a topological space - since the properties of open sets in a metric space inspired the general definition of topology. The topology arising from the usual metric on R^n is called the *usual topology*.

2. If S is any set then the family of *all* subsets of S naturally satisfies (1), (2) and (3) and so is itself a topology. It is called the *discrete topology*. It is the topology obtained by giving S the discrete metric (§1.2, Example 8), as follows easily from the observation that each point by itself forms an open set with respect to the discrete metric because $B_\delta(x)$ consists of the point x alone when $\delta < 1$.

3. At the other extreme, we may take the family consisting of *no* open sets except those legally required by clause (3), namely S itself and the empty set $\emptyset$. This is sometimes called the *indiscrete* topology.

4. Suppose S is a set with a given topology F, and T is some arbitrary subset of S (not necessarily belonging to F). Then T can

be given a topology F_T as follows: V belongs to F_T precisely when V is of the form $T \cap U$ where U is some set belonging to F. The verification of (1), (2), (3) for F_T is easy. This topology F_T is called the *induced topology* or *subspace topology* on T. If F comes from a metric on S then F_T in fact comes from the induced metric (§1.2, Example 7) on T.

5. If S, T are two topological spaces there is a natural way of constructing a topology on the cartesian product $S \times T$ from those of S and T. It is tempting to define a set in $S \times T$ to be open if it is of the form $U \times V$ where U, V are open in S, T respectively, but unfortunately this fails to satisfy rule (2) in general. (For example if $S = T = R$ with the usual topology then sets $U \times V$ can be thought of as rectangles, and a union of rectangles may not be a rectangle.) Instead we define a set in $S \times T$ to be *open* if it is the *union of sets of the form* $U \times V$ *where* U, V *are open in* S, T *respectively*. It is straightforward to verify that this does satisfy the rules for a topology on $S \times T$. It is called the *product topology* on $S \times T$.

It is an instructive exercise to check that the product topology on $R^n \times R^m$ (R^n, R^m with usual topology) is the same as the usual topology on $R^{n+m} = R^n \times R^m$.

6. Let S be a topological space, and suppose that S is expressed as the disjoint union of a (possibly infinite) family of sets S_λ. For example, in certain types of application the elements of S may be some objects which we are attempting to classify and each S_λ may represent a collection of objects having a certain property P_λ in common. If we now broaden our focus and decide to regard two objects as 'the same' if they belong to the same S_λ, we are led to consider instead of S a

new set $\tilde{S}$, namely the set whose elements are themselves the S_λ. This set $\tilde{S}$ inherits a topology from that of S in the following way. Given any set $\tilde{W} \subset \tilde{S}$ let W denote the subset of S consisting of the union of all those S_λ which belong to $\tilde{W}$. Then define $\tilde{W}$ to be *open* in $\tilde{S}$ if W is open in S. See Figure 7. Again,

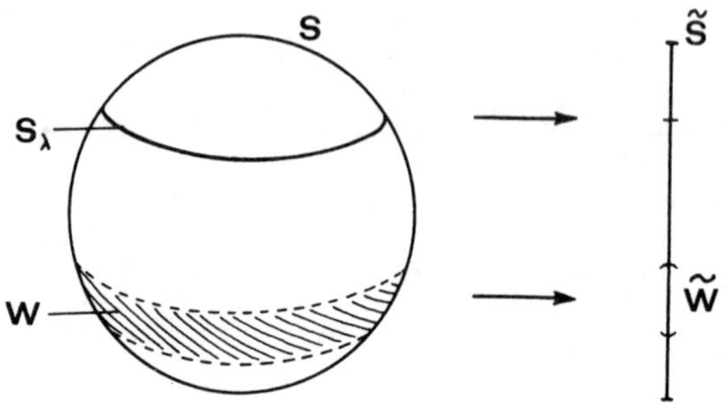

Figure 7

it is routine to verify that this defines a topology on $\tilde{S}$.

There is an equivalent but neater way to express this construction. Let S be a topological space and let R be some equivalence relation defined on S. The equivalence classes of R give a decomposition of S into disjoint subsets S_λ as above, and conversely such a decomposition defines an equivalence relation on S - namely, that of belonging to the same S_λ. Denote the set of equivalence classes by S/R: this corresponds to $\tilde{S}$. We have a natural map

$$\pi : S \to S/R$$

taking each $x \in S$ to its equivalence class. Define a set $U \subset S/R$ to be *open* if $\pi^{-1}(U)$ is open in S. Then this is the same definition as above, and so it equips S/R with a topology. This topology on S/R $(= \tilde{S})$ is called the *quotient topology*.

If the topology on S comes from a metric it will not necessarily follow that the quotient topology comes from a metric on

S/R: there is no such thing as a 'quotient metric' in general. This is one more reason for studying topologies rather than metrics.

Returning now to the motivating idea behind the generalization of metric space to topological space, we will formulate a definition of *continuity* for maps between topological spaces. In view of the above theorem and the subsequent discussion there is only one plausible definition that coincides with our previous definition for metric spaces.

DEFINITION

Let $f : A \to B$ *be a map between two topological spaces (topologies understood). Then* f *is* continuous *if* $f^{-1}(V)$ *is open in* A *whenever* V *is open in* B.

Again we emphasize that there is nothing here to say that if U is open in A then f(U) should be open in B.

The main advantage of this topological definition of continuity is that it avoids reference to any specific metric, whether or not there is one available. It is often useful to work with topological spaces (such as function spaces) which can be given metrics but only in rather artificial ways that distract attention from the more natural topological structures. Indeed, there are contexts in which it is necessary to deal with topological spaces having no metric structure at all which is compatible with the topology.

The second advantage is the usefulness and economy of the definition for handling theoretical statements about continuous maps. For example, consider the following proof of the fact that the composition of two continuous maps is continuous. The proof for metric spaces (using δ, ε) is of course straightforward, but not as tidy.

THEOREM

If $f : A \to B$ *and* $g : B \to C$ *are continuous then so is the composition* $g \cdot f : A \to C$.

Proof

Choose any open set V in C. Then $g^{-1}(V)$ is open in B by the continuity of g, so $f^{-1}(g^{-1}(V))$ is open in A by the continuity of f. But $f^{-1}(g^{-1}(V)) = (g \cdot f)^{-1}(V)$; this finishes the proof.

1.4 FURTHER TOPOLOGICAL CONCEPTS

(i) An open set U containing the point x is called an *open neighbourhood* of x. For technical reasons it can be useful to think in slightly more general terms, and to call any set N simply a *neighbourhood* of x if it contains an open neighbourhood of x. Thus for example the interval $\{x \in \mathbb{R} \mid 1 \leqslant x \leqslant 3\}$ is a neighbourhood (though not open) of the point 2 in $\mathbb{R}$, but it is not a neighbourhood of either of the points 1 or 3 in $\mathbb{R}$.

There is a characterization of open sets that is convenient to use in practice when checking whether a given set is open or not. The proof is straightforward from the definition of topology (page 14) with clause (2) playing an important part.

Characterization of open sets

A set U is open if and only if for each point $x \in U$ there is a neighbourhood N_x of x with N_x contained in U.

We see from this that we can use 'neighbourhood' in topological spaces in contexts where we would use B_ε in metric spaces. For example, we can define a map $f : A \to B$ to be *continuous at a point* $x_o \in A$ ·

(a concept which we have only used so far in metric spaces) if *given any neighbourhood* N *of the point* $y_o = f(x_o) \in B$ *there exists a neighbourhood* M *of* x_o *with* $f(M) \subset N$. It is now a routine matter to check, using the above characterization of open sets, that a map $f : A \to B$ between two topological spaces is continuous (as defined on page 17) if and only if it is continuous at x_o (as defined above) for every $x_o \in A$. This now brings us back satisfyingly in a full circle to the original definition of continuity in a metric space (page 9) with which we began.

(ii) Given any set W in a topological space S we can consider *the union of all the subsets of* W *which are open sets in* S (so open in W with the induced topology - but the converse may not hold). This is open (by clause (2)) and so can be thought of as *the largest open set contained in* W. It is called the *interior* of W, and in a sense measures how 'thick' W is as a subset of S. Thus a line or a plane has empty interior in R^3. The interior of the closed interval $[a,b]$ in R is the open interval (a,b), but both have empty interior in the plane R^2. The set Q of rational numbers and its complement I, the set of irrational numbers, each have empty interior in R.

(iii) A subset of a topological space is called a *closed* set if its *complement is an open set*. Generally speaking, most sets are neither open nor closed; it is also possible for a set to be both open and closed simultaneously. In a discrete space, for example, all sets are both open and closed. Thus a closed set is *not* simply one which fails to be open. The reason for introducing this terminology is that topological arguments frequently go through much more rapidly when expressed in terms of closed sets than they would do if they dealt with open sets directly.

Observe that by taking complements the useful characterization of open sets given above is transformed automatically into the following:

Characterization of closed sets

A set C is closed if and only if for each point y not in C there is a neighbourhood N_y of y entirely disjoint from C.

(iv) Let W be any subset of the topological space S, and define a point x (which may or may not belong to W) to be an *accumulation point* or *limit point* of W if *each neighbourhood of* x *contains at least one point of* W, *apart from (possibly) the point* x *itself*. Thus there are points of W 'arbitrarily close' to x but different from x. It is easy to see that the characterization of closed sets above can now be converted into the statement that *a set is closed if and only if it contains all its limit points*.

Given an arbitrary set W, not necessarily closed or open, we can manufacture a set $\bar{W}$ consisting of W together with all its limit points. This set $\bar{W}$ is called the *closure* of W. Obviously $W \subset \bar{W}$, and the set C is closed if and only if $C = \bar{C}$. The following facts are not difficult to verify:

(1) For any set W the closure $\bar{W}$ is a closed set, i.e. $\bar{\bar{W}} = \bar{W}$.

(2) $\bar{W}$ is the intersection of all closed sets that contain W.

(3) $\overline{W_1 \cup W_2} = \bar{W}_1 \cup \bar{W}_2$.

However, note that $\overline{W_1 \cap W_2} \neq \bar{W}_1 \cap \bar{W}_2$ necessarily. For example, take $W_1 = \{x \in R \mid x < a\}$, $W_2 = \{x \in R \mid x > a\}$. Then $\overline{W_1 \cap W_2} = W_1 \cap W_2 = \emptyset$ but $\bar{W}_1 \cap \bar{W}_2 = \{a\}$.

If A is a subset of B and yet occupies so much of B that $\bar{A}$ is equal to or contains B, then we say that A is *dense* in B. Other equivalent ways of saying this are that every point of B is either in A or is an accumulation point of A, or that every open subset of B contains points of A, or that the complement of A in B has empty interior. For example, the set of non-integer numbers is dense (and open) in $\mathbb{R}$. The set of rational numbers is dense (and not open) in $\mathbb{R}$. The concept of denseness is very important in mathematical characterizations of 'usual' as opposed to 'unusual' properties of maps, dynamical systems, and so on: we shall make considerable use of it. See §3.3 (Sard's theorem) and also §4.4.

(v) The familiar ideas of *approaching a limit* and *convergence* can easily be fitted into the topological framework constructed so far. In a metric space context (e.g. in euclidean space) we say

> $f(x)$ *approaches* c *as* x *approaches* x_0 if, given any $\varepsilon > 0$, there then exists some $\delta > 0$ with the property that $d(f(x), c) < \varepsilon$ for all x with $d(x, x_0) < \delta$.

In topological terms this becomes

> $f(x)$ *approaches* c *as* x *approaches* x_0 if, given any neighbourhood N of c there then exists some neighbourhood M of x_0 with the property that $f(x) \in N$ for all $x \in M$.

We write

$$f(x) \to c \quad \text{as} \quad x \to x_0$$

or

$$\lim_{x \to x_o} f(x) = c$$

and call c the *limit* of $f(x)$ as $x \to x_o$.

The wording of these definitions sounds familiar, and in fact we note that it is an echo of the definition of 'continuity at a point' given earlier in this section. What, then, is the exact relationship between these ideas? Proceeding carefully from the definitions alone we soon see that to say

$$f(x) \to c \quad \text{as} \quad x \to x_o$$

is the same as to say that the map g defined by

$$\left.\begin{aligned} g(x) &= f(x) \quad \text{when} \quad x \neq x_o \\ g(x_o) &= c \end{aligned}\right\}$$

is continuous at x_o. Therefore in making the definition of 'approaching a limit' we have actually introduced nothing new.

A standard variant on the above occurs when we are dealing with a *sequence* of points $x_1, x_2, x_3, \ldots$. In this case we say

$$x_n \to c \quad \text{as} \quad n \to \infty$$

or

$$\lim_{n \to \infty} x_n = c$$

to mean that

> given any neighbourhood N of c there then
> exists some positive integer m such that
> $x_n \in N$ for all $n > m$.

For metric spaces (in particular euclidean space) this coincides with the usual idea of a sequence approaching a limit: given any $\varepsilon > 0$ there exists some positive integer m such that $d(x_n,c)$ is less than ε for all $n > m$.

Here, and in the general topological setting, we say that the sequence is *convergent*, or *converges* to the limit c.

In the special case of real or complex-valued functions or sequences the definitions can be stretched a little to include the notion of 'approaching infinity'. We replace the neighbourhood N of c by a set of the form $\{y \in \mathbb{R} \mid y > r\}$ for some r when defining $f(x) \to +\infty$ (or $x_n \to +\infty$) in the real-valued case, or by a set of the form $\{y \in \mathbb{R} \mid y < s\}$ for some s when defining $f(x) \to -\infty$ (or $x_n \to -\infty$) also in the real case, or, finally, by a set of the form $\{z \in \mathbb{C} \mid |z| > r\}$ to define $f(x) \to \infty$ (or $z_n \to \infty$) in the complex-valued case.

Remarks

1. The definitions for sequences can actually be subsumed under those for functions. The way to do this is to consider the sequence $x_1, x_2, x_3, \ldots$ of points in the space S as the images of the integers $1, 2, 3, \ldots$ under the map $\phi : \mathbb{N} \to S$ taking 1 to x_1, 2 to x_2, and so on. By adjoining an extra point p (thought of as 'infinity') to $\mathbb{N}$ and putting a suitable topology on $\mathbb{N} \cup \{p\}$ it is possible to re-express '$x_n \to c$ as $n \to \infty$' as

$$\phi(n) \to c \quad \text{as} \quad n \to p .$$

We will not go into details, since they are rather uninstructive.

2. There is an irritating source of ambiguity which inevitably arises at this point. We have the notion of *limit* of a sequence and *limit point* of a set, but when we think of a sequence as a set of points these two

notions may not coincide. For example, the sequence $-\frac{1}{2}, +\frac{3}{4}, -\frac{5}{6}, +\frac{7}{8}, \ldots$ $\left(\text{i.e. } x_n = (-1)^n \frac{2n-1}{2n}\right)$ has no limit, although the points ± 1 are both limit points of the set $\{x_n \mid n \in N\}$. On the other hand, the sequence 1, 1, 1, 1, ... has the limit 1 but as a *set* it has no limit points since it consists of only one point. Therefore care must be taken to use these terms with precision when sequences are being treated from a topological point of view.

3. When S is a topological space which also carries some notion of *addition* (for example a normed linear space: see §2.3), then besides the concept of convergent *sequence* we also have that of convergent *series*. A series means simply a sequence of elements of S with '+' signs between them:

$$x_1 + x_2 + x_3 + \ldots \quad .$$

We say the series is *convergent* or otherwise according to whether the sequence

$$S_1, S_2, S_3, \ldots$$

is convergent or otherwise, where the S_n are the so-called *partial sums*

$$S_n = \sum_{i=1}^{n} x_i \ .$$

We emphasize, however, that this only makes sense to the extent that '+' makes sense in S. We write $\sum_{i=0}^{\infty} x_i$ to denote $\lim_{n\to\infty} S_n$.

4. Any metric space S (such as R^n) has the elementary property that *given* $x, y \in S$ *there exist neighbourhoods* U, V *of* x, y *respectively such that* U *and* V *are disjoint*. This may seem obvious, but must be verified from the definitions. In fact we may take $U = B_{d/2}(x)$ and $V = B_{d/2}(y)$ where $d = d(x,y)$: then the non-intersection of U, V

follows from the triangle inequality. Unfortunately topological spaces failing to have this 'separation' property can arise from quite natural considerations (e.g. in quotient spaces: see page 16, and §4.1). Spaces which *do* have the property are called *Hausdorff* spaces (F. Hausdorff, 1868-1942). Thus all metric spaces are Hausdorff.

1.5 HOMEOMORPHISM OF SPACES AND EQUIVALENCE OF MAPS

Let S, T be two topological spaces, and suppose $f : S \to T$ is a bijection*. If f is continuous, and at the same time its inverse $f^{-1} : T \to S$ is continuous, then f is called a *homeomorphism*. From the definition of continuity, this means that the open sets in S correspond precisely to the open sets in T, under f. Thus if there exists a homeomorphism $f : S \to T$ then as far as their topological structure is concerned S and T are indistinguishable. We say that S and T are *topologically equivalent* or, more usually, *homeomorphic*. It follows that any property possessed by S and expressible purely in terms of the topology of S (such as the property of being Hausdorff, for example) will also be possessed by T, and vice versa.

In many useful contexts the fact that a bijection f is continuous guarantees that its inverse f^{-1} is continuous: see the following section for a particular theorem on these lines. This means that it is then not necessary to study f^{-1} explicitly in order to show that f is a homeomorphism, which is very useful since f^{-1} may in practice be difficult to analyze. Unfortunately this result does not hold in all generality. It may fail, for example, in certain function spaces. Artificial counterexamples are also easily manufactured. Let $S = \mathrm{R}$ with the discrete topology and $T = \mathrm{R}$ with the usual topology, and let

* see page 271.

$f : S \to T$ be any bijection: then f is continuous (*every* map with domain S is continuous!) but f^{-1} is not since there exist open sets U in S with fU not open in T.

Homeomorphism is the natural notion of equivalence between two topological spaces. We next consider how to define a notion of equivalence between two maps $f,g : S \to T$ where the two topological spaces S, T are given. Thinking informally of a homeomorphism from a space S to itself as the result of viewing S through distorting spectacles, we can regard two maps, $f,g : S \to T$ as being equivalent if g looks like f when S and T are both viewed through distorting spectacles - but with separate lenses. Formally this is expressed by saying that f, g are *topologically equivalent* or *conjugate* if there exist homeomorphisms

$$h : S \to S \qquad\qquad k : T \to T$$

such that

$$g \cdot h = k \cdot f \quad ,$$

i.e. the result of applying g to S after distortion by h is the same as the result of applying f to S and then distorting the result by k. Note that this is the same as

$$g = k \cdot f \cdot h^{-1} \quad ,$$

and can be expressed simply by saying that the following diagram

$$\begin{array}{ccc} S & \xrightarrow{f} & T \\ h \downarrow & & \downarrow k \\ S & \xrightarrow{g} & T \end{array}$$

commutes, i.e. either pathway from the top left hand corner to the bottom right hand corner gives the same result. Observe that since h, k have inverses the relation of topological equivalence is an equivalence relation on the set of continuous maps $S \to T$.

In cases when $S = T$ we often insist that k should be the same as h, since we wish to avoid the idea of distorting S in two different ways simultaneously. The expression for topological equivalence then becomes

$$g = h \cdot f \cdot h^{-1} \quad .$$

We shall see more of this later, when studying the qualitative theory of dynamical systems (§§4.2, 4.5).

1.6 COMPACTNESS

There is an extremely useful theorem in analysis as follows:

THEOREM A

If f *is a continuous real-valued function on a closed interval* $[a,b]$ *then* f *is bounded, i.e. there exists a constant* k *such that* $|f(x)| \leq k$ *for all* x *in* $[a,b]$.

Note that the assumption that $[a,b]$ is a *closed* interval is vital. For example, $\tan x$ is defined and continuous on the non-closed interval $[0, \frac{\pi}{2})$ but is certainly not bounded.

Here are two standard proofs of the theorem.

Proof 1

Suppose f is *not* bounded: then we deduce a contradiction. There must exist $x_1 \in [a,b]$ with $|f(x_1)| > 1$ (otherwise f is bounded by 1). Then there exists $x_2 \neq x_1$ with $|f(x_2)| > 2$ (otherwise f is bounded by the greater of 2 and $|f(x_1)|$). Continuing, we obtain distinct points $x_1, x_2, x_3, \ldots$ with $|f(x_n)| > n$ for each positive integer n.

Now we invoke the 'classical' *Weierstrass limit point theorem* (or *condensation theorem*):

W *If* A *is a subset of a closed interval and* A *contains an infinite number of points then there is at least one accumulation point of* A *somewhere in the interval.*

This immediately implies that there is in $[a,b]$ an accumulation point x_* of $A = \{x_1, x_2, x_3, \ldots\}$, i.e. every $B_\delta(x_*)$ contains some x_n. But by the continuity of f at x_* we see (taking $\varepsilon = 1$) that there is a $\delta_1 > 0$ such that $|f(x)| < |f(x_*)| + 1$ for all x in $B_{\delta_1}(x_*)$. Choose now m to be an integer bigger than $|f(x_*)| + 1$, and choose $\delta \leqslant \delta_1$ so small that $B_\delta(x_*)$ avoids $x_1, x_2, \ldots, x_{m-1}$. Then $B_\delta(x_*)$ contains some x_n with $n \geqslant m$, and $|f(x_n)| > n \geqslant m > |f(x_*)| + 1$ which gives the contradiction.
(Note: we do not claim that $x_n \to x_*$.)

Proof 2

Choose $\varepsilon > 0$. By the continuity of f we know that for each x in $[a,b]$ there exists $\delta_x > 0$ such that $|f(y)|$ is less than $|f(x)| + \varepsilon$ provided y is within δ_x of x. The whole of $[a,b]$ is covered by the open intervals $(x - \delta_x, x + \delta_x)$ as x runs through $[a,b]$, or equivalently

$$[a,b] = \bigcup_{x \in [a,b]} (x - \delta_x, x + \delta_x)'$$

where the ' denotes that part of $(x - \delta_x, x + \delta_x)$ lying in $[a,b]$. (Note in passing that $(x - \delta_x, x + \delta_x)'$ is an open subset of $[a,b]$ in the induced topology of $[a,b]$ as a subspace of R.) This time we use the *Heine-Borel theorem*:

HB *Suppose* $[a,b]$ *is a closed interval in* R *(with* $a,b \neq \pm\infty$*) and there is given a family* $\{J_\lambda\}$ *of* open *intervals covering* $[a,b]$*, i.e.* $[a,b] \subset \bigcup_\lambda J_\lambda$. *Then there is a* finite sub-family $J_{\lambda_1}, J_{\lambda_2}, \ldots, J_{\lambda_m}$ *which also covers* $[a,b]$*, i.e.* $[a,b] \subset J_{\lambda_1} \cup J_{\lambda_2} \cup \ldots \cup J_{\lambda_m}$.

Applying this to the family $\{J_x\}$ where $J_x = (x - \delta_x, x + \delta_x)$ we deduce that there exist $x_1, x_2, \ldots, x_m$ in $[a,b]$ with $[a,b] \subset J_{x_1} \cup J_{x_2} \cup \ldots \cup J_{x_m}$. Then clearly $|f(x)|$ is everywhere less than $\max\{|f(x_1)|, |f(x_2)|, \ldots, |f(x_m)|\} + \varepsilon$ which shows that f is bounded.

It is straightforward to follow through the above two proofs and check that they work for functions defined on any topological space S which satisfies an analogue of W or HB respectively where the closed interval $[a,b]$ is replaced by S and *open interval* is replaced by *open set of* S in HB . In fact topological spaces that satisfy this analogue of the Heine-Borel theorem play such a fundamental role in analysis that they are given an explicit name: *compact* spaces.

DEFINITION

A topological space S *is* compact *if and only if every cover of* S *by open sets has a* finite *subcover.*

Sometimes it is convenient to talk about a compact subset T of a space S. This simply means that T is compact when regarded as a topological space in its own right (with the induced topology).

In the present more general context it is easy to deduce W for compact spaces as a consequence of HB . The argument goes as follows.

Suppose A is a subset of the compact topological space S and A has an infinite number of points. We wish to prove A has an accumulation point. If not, then every point of S has an open neighbourhood containing at most *one* point of A. But then the compactness of S implies that a finite number of these open neighbourhoods will suffice to cover S, which means that A can have only a finite number of points. This contradiction shows that A must after all have an accumulation point.

Returning to the starting-point of this section, by mimicking either Proof 1 (using the above generalized version of $\underline{W}$) or Proof 2 directly we can construct a proof of a more general version of the theorem on boundedness of continuous functions.

THEOREM A'

If S *is a compact topological space and* $f : S \to R$ *is continuous then* f *is* bounded, *i.e. there exists a constant* k *such that* $|f(x)| \leq k$ *for all* $x \in S$.

Remarks

1. Unfortunately it is not possible to deduce compactness from the Weierstrass property in the case of an otherwise arbitrary topological space. Thus the generalized versions of $\underline{W}$ and $\underline{HB}$ are not entirely equivalent. In the case of *metric spaces*, however, they can be shown to be equivalent. In particular they are equivalent in their original context, i.e. for closed intervals $[a,b]$ of the real line.

2. Theorem A' has an immediate corollary which illustrates one of the main ways in which the theorem might be used in practice, namely that if S is compact and $f : S \to R$ is a continuous non-vanishing function then f

is *bounded away from zero*, i.e. there exists $k > 0$ such that $|f(x)| \geq k$ for all $x \in S$. The proof of the corollary comes from applying Theorem A' to $1/f$. Clearly this need not be true if S is not compact: take $S = (0,1)$ and $f(x) = x$, for example. In applications f may typically be a distance function from points of S to some other point or set.

3. There is an alternative characterization of compactness as follows. A topological space S is compact if and only if, *given any family of closed subsets* $\{C_\lambda\}$ *such that any* finite *number of the* C_λ *have non-empty intersection, then the intersection of them* all *is non-empty.* At first sight this seems unrelated to the former definition, but in fact it is easy to translate one version into the other simply by considering the open complements U_λ of the closed sets C_λ, and vice versa. Thus for example the statement $S = \bigcup_\lambda U_\lambda$ becomes $\emptyset = \bigcap_\lambda C_\lambda$, and so on. We leave this verification as an exercise.

This characterization is frequently used to prove the existence of a point defined by some kind of infinite intersection process. As an illustration, observe that if I_n is the closed interval $[0, \frac{1}{n}]$ then the intersection of any finite number of I_n's is non-empty and in fact $\bigcap_{n=1}^{\infty} I_n = \{0\}$ which is non-empty. On the other hand if $J_n = (0, \frac{1}{n}]$ then again the intersection of any *finite* number of J_n's is non-empty but now $\bigcap_{n=1}^{\infty} J_n = \emptyset$. Here $J_1 = (0,1]$ is not compact (see below).

We will now give a working theorem whose geometrical usefulness justifies the rather technical definition of compactness. First we state three *facts about compact sets*, each of which is straightforward to verify from the definitions:

(1) Every closed subset of a compact space is compact.

(2) If $f : S \to T$ is continuous and $K \subset S$ is a compact subset, then $f(K)$ is a compact subset of T.

(3) In a Hausdorff space every compact set is closed.

Putting these together we obtain

THEOREM B

Let S *and* T *be topological spaces with* S *compact and* T *Hausdorff. Suppose* $f : S \to T$ *is a continuous map which is also a bijection. Then* f *is automatically a homeomorphism (i.e.* f^{-1} *is automatically continuous).*

Proof

To prove f^{-1} continuous we have to show $(f^{-1})^{-1}(V)$ is open whenever V is, i.e. V open implies $f(V)$ open. This is the same (since f is a bijection) as showing C closed implies $f(C)$ closed. But C closed implies C compact (Fact (1)) which implies $f(C)$ compact (Fact (2)) which in turn implies $f(C)$ closed (Fact (3)).

Thus for example to prove that the space of real numbers modulo the integers (i.e. the quotient space R/R where xRy means $x - y \in Z$) is formally homeomorphic to a circle it suffices to construct a continuous bijection f from R/R to the unit circle S^1 in the complex plane, since R/R is compact (being the image of $[0,1]$ under the projection $R \to R/R$) and S^1 is Hausdorff. It is easy to verify that f : (equivalence class of $\theta) \mapsto e^{2\pi i\theta}$ is a bijection, and is continuous by definition of the quotient topology in R/R and the continuity of the map $R \to S^1 : \theta \mapsto e^{2\pi i\theta}$. We have no need to bother with the technicalities of working with f^{-1}.

Recognizing compact sets. In all the above theory we have not discussed how to recognize a compact set where we see one. After all, it is clearly impossible to check the properties of every open cover. Fortunately, in euclidean space $\mathbb{R}^n$, the most useful case, there is an immediate test:

THEOREM C

A subset K *of* $\mathbb{R}^n$ *is compact precisely when it is both (a) closed and (b) bounded (i.e. there exists some number* k *with* $K \subset B_k(0)$*).*

The compactness of K immediately implies (a) (Fact (3)) and (b) (cover K by $K \cap B_k(0)$, $k = 1, 2, 3, \ldots$). However, the converse is more delicate, being a generalized Heine-Borel theorem. We will not give the proof here. Note that once we have Theorem C then the original Theorem A follows directly from Fact (2), since to say that f is bounded on $[a,b]$ is the same as saying $f([a,b])$ is a bounded subset of $\mathbb{R}$.

1.7 CONNECTEDNESS

A topological space is *connected* if, roughly speaking, it is all in one piece. Of course, any space with more than one point in it can be written as the union of two disjoint non-empty subsets, but it turns out that the definition which best captures our intuitive notion of being 'in one piece' is the following:

DEFINITION

A topological space S *is* connected *if it cannot be split into two disjoint non-empty subsets which are both open, i.e. if* $S = U \cup V$ *where* U, V *are open and disjoint then one of them must be empty.*

Thus in a connected space the only subsets which are both open and closed simultaneously are the empty set and the whole space. It follows that any space with the discrete topology cannot be connected if it has more than one point. A subset Q of R^n, when given the induced topology, will fail to be connected precisely when it can be written as $Q = Q_1 \cup Q_2$ where Q_1, Q_2 are non-empty, disjoint, and $Q_i = Q \cap U_i$ for some open set U_i of R^n - in other words when each Q_i can be included in an *open* set U_i which does not meet the other Q_i. This formalizes the notion of Q_i being 'separate pieces'. For example, the union Q of the two closed intervals $Q_1 = [0,1]$ and $Q_2 = [2,3]$ is *not* connected (as a subset of R with the induced topology), since we have $Q_1 = Q \cap (-2,2)$ and $Q_2 = Q \cap (1,4)$.

This still leaves the problem of deciding when a given topological space *is* connected, since it is impossible to check all possible attempts at splitting it into disjoint open subsets. The two most useful results in this direction are:

(1) If S is connected and $f : S \to T$ is continuous then $f(S)$ (with the topology induced from T) is connected;

(2) every open, half-open or closed interval (possibly infinite) of the real line is connected.

Sketch proofs

(1) If $f(S) = U \cup V$ (both open and disjoint in $f(S)$) then $S = f^{-1}(U) \cup f^{-1}(V)$, both open in S by the continuity of f and clearly disjoint.

(2) If an interval I is *not* connected we have $I = U \cup V$, both open in I, disjoint and non-empty. Choose points a in U, b in V, suppose $a < b$, and let m = least upper bound of the set X of all numbers x for which $[a,x) \cap I$ is contained in U. Contemplation of m shows m cannot lie in U (since it would not be an upper bound for X) or V (since it would not be the least upper bound for X). Contradiction, so I *is* connected.

As with compactness, we extend the definition to *subsets* of a topological space by regarding them as spaces in their own right with the induced topology.

Apart from its direct geometrical meaning, connectedness plays an important indirect role in the proofs of many theorems. Suppose we wish to prove that a certain property holds for all x in some topological space S. A frequently useful approach is to prove first of all that the set T of points x for which P does hold is an open subset of S, and then to prove by other means that the set F of points for which P fails is also an open subset of S. If S is connected, then the existence of just one point x for which P holds means that T is non-empty, so F must be empty and therefore the validity of P is guaranteed for all x in S.

Remarks on the literature. There are many good introductory texts on metric and topological spaces, such as Mendelson [80], Patterson [94], Pitts [99]. Particularly recommended also are Sutherland [134], which is written in much the same spirit as this chapter but with all details provided, and Simmons [118], which contains much additional material relevant to the notes as a whole. See Willard [150] for a deeper study of topological spaces, or Hocking and Young [57] for more geometric aspects.

2 Calculus

2.1 DIFFERENTIATION

Recall from elementary calculus the definition of *differentiability* at a point x for a real-valued function of one real variable.

DEFINITION

The function $f : R \to R$ *is* differentiable *at* x *if*

$$\lim_{h \to 0} \frac{1}{h} (f(x+h) - f(x))$$

exists ($\pm \infty$ *not admitted*).

This limit is called the *derivative* of f at x, and is traditionally denoted by $f'(x)$ or $\frac{df}{dx}$. Strictly speaking we should write $\frac{df}{dx}(x)$, to emphasize the point x for which the limit has been evaluated.

Another way of expressing the existence of the limit is to say:

there exists a real number λ *such that*

$$f(x+h) - f(x) = \lambda h + \varepsilon|h|$$

where $\varepsilon \to 0$ *as* $h \to 0$.

In general. λ of course depends on x. The pictorial interpretation is that the graphs of the two functions $\bar{f}$, L of h defined by

$$\bar{f}(h) = f(x+h) - f(x)$$

$$L(h) = \lambda h$$

are *tangent* at $h = 0$: see Figure 8.

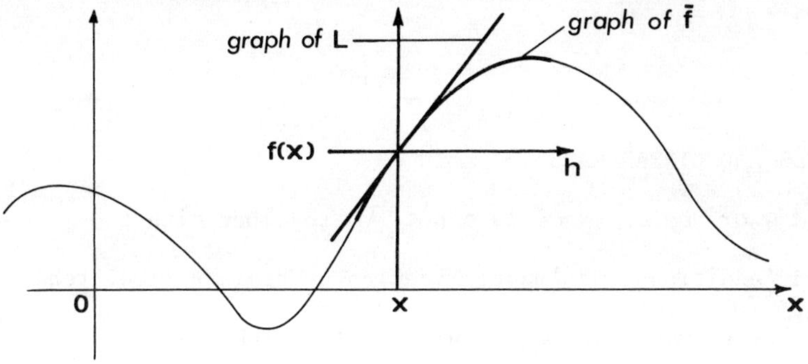

Figure 8

It is reasonable to describe this by saying that L is a *linear approximation* to f at x, or, equivalently, that the map

$$h \mapsto f(x) + \lambda h$$

(whose graph is a straight line) is a *first order approximation* to the map

$$h \mapsto f(x+h) \quad .$$

Since this idea turns out to be so useful in the study of real functions of a real variable, it is worthwhile trying to generalize it as much as possible. Obviously we would get nowhere by considering functions (maps) between arbitrary topological spaces, since the minimum necessary ingredients are:

1. concepts of *addition and subtraction*,
2. the idea of a map being *linear*,
3. a possibility of taking *limits*.

Now the first two of these are available in the world of linear spaces or vector spaces, while the third is most conveniently associated with metric spaces (§1.2). Therefore to generalize our simple notion of differentiation we need to develop a theory of linear spaces which are at the same time metric spaces. This will occupy us for the next two sections of the chapter.

2.2 LINEAR SPACES AND LINEAR MAPS

A *real linear space* or *vector space* is a set V together with two operations called 'addition' and 'multiplication by scalars (i.e. real numbers)' which obey a number of reasonable rules. The standard rules for addition are these:

(i) $x + (y+z) = (x+y) + z$ for all x, y, z in V

(ii) $x + y = y + x$ for all x, y in V

(iii) there is a *zero* in V, i.e. an element 0 which satisfies $0 + x = x + 0 = x$ for every x in V

(iv) every x has a *negative*, i.e. an element $-x$ which satisfies $x + (-x) = 0$,

followed by rules involving scalar multiplication as well as addition:

(v) $\alpha(x+y) = \alpha x + \alpha y$ for all x, y in V and all scalars α

(vi) $(\alpha+\beta)x = \alpha x + \beta x$ for all x in V and all scalars α, β

(vii) $\alpha(\beta x) = (\alpha\beta)x$ for all x in V and all scalars α, β

(viii) $1x = x$ for all x in V.

Remarks

1. It can be *deduced* from these rules that other obvious-looking arithmetical statements also hold, such as $0x = 0$ or $(-1)x = -x$. Also, it is usual to write $x - y$ instead of $x + (-y)$.

2. The same definition could equally well be made with complex numbers rather than real numbers as scalars, giving a *complex linear space*.

3. When thinking geometrically we call 0 the *origin* of V.

4. Nothing has been said about any topology for V.

EXAMPLES of linear spaces

1. $V = R$; usual addition and multiplication.

2. $V = R^n$; co-ordinate-wise addition and scalar multiplication.

3. $V = \{\text{all functions } R \to R\}$; addition and scalar multiplication defined *pointwise*, i.e.

$f + g$ means the function $x \mapsto f(x) + g(x)$

αf means the function $x \mapsto \alpha f(x)$.

More generally:

4. If W is a linear space and A is *any* set whatsoever then

$$V = \{\text{all maps } A \to W\}$$

is a linear space with pointwise addition and scalar multiplication. The linear structure in W is all that is needed to make this definition work.

5. $V = \{\text{all solutions } x(t) \text{ of } \ddot{x} + x = 0\}$; pointwise addition and scalar multiplication.

6. $V = \{0\}$. (Observe that $V = \emptyset$ is *not* a linear space, since it does not satisfy rule (iii).)

7. If S is a subset of a linear space V, and *all sums and scalar multiples of elements of* S *are also in* S, then S automatically becomes a linear space in its own right. It is called a *linear subspace* of V. Here are some examples:

(a) $V = R^3$, S = any line or plane through the origin.

(b) V = Example 3 above,

(i) S = all functions differentiable at every point,

(ii) S = Example 5.

(c) V = Example 5 above, S = all $x(t)$ with $x(t_0) = 0$ for some given t_0. (Note that if we replaced 0 by something non-zero in this example then S would no longer be a linear subspace since the sum of two elements of S would not lie in S.)

(d) If S, T are linear subspaces of V then so are $S \cap T$ and the set $S + T$ of all sums $s + t$ where $s \in S$, $t \in T$. Note that $S \cup T$ in general is not a linear subspace of V.

8. If V, W are linear spaces then so is the cartesian product $V \times W$ with co-ordinate-wise addition and scalar multiplication. Note that $\{0\} \times W$ and $V \times \{0\}$ are linear subspaces of $V \times W$.

In working with linear spaces it is natural to pay particular attention to those maps between linear spaces which respect the linear structures. Such maps are called *linear maps*.

DEFINITION

Let V, W *be linear spaces. A map* $L : V \to W$ *is a* linear map *if*

$$L(\alpha u + \beta v) = \alpha L(u) + \beta L(v)$$

for every u, v *in* V *and all scalars* α, β.

Note here that the '+' on the left hand side denotes addition in V, whereas on the right hand side it denotes addition in W. These may be completely different in the way they are defined. A similar remark applies to the scalar multiplications on both sides. Observe that a linear map always takes zero (in V) to zero (in W).

EXAMPLES of linear maps

1. V = any linear space; $L : V \to V$ defined by $L(v) = \alpha v$ for some fixed scalar α.

2. $V = \mathbb{R}^n$, $W = \mathbb{R}^m$; $L : V \to W$ defined by

$$L(x_1, x_2, \ldots, x_n) = \left(\sum_{j=1}^{n} \alpha_{1j}\, x_j,\ \sum_{j=1}^{n} \alpha_{2j}\, x_j,\ \ldots,\ \sum_{j=1}^{n} \alpha_{mj}\, x_j\right)$$

for a collection of mn fixed scalars α_{ij} $(1 \leqslant i \leqslant m,\ 1 \leqslant j \leqslant n)$. Here L is traditionally represented by the *matrix*

$$\begin{pmatrix} \alpha_{11} & \alpha_{12} & \cdots & \alpha_{1n} \\ \alpha_{21} & \alpha_{22} & \cdots & \alpha_{2n} \\ \vdots & \vdots & & \vdots \\ \alpha_{m1} & \alpha_{m2} & \cdots & \alpha_{mn} \end{pmatrix} .$$

3. $V = S$ in Example 7(b) (i) above, $W = V$ in Example 3 above; $L : V \to W$ defined by $L(f) = f'$.

4. Linear maps from V to W can be added together and multiplied by scalars pointwise to give further linear maps. Indeed, it is easy to check that the set $\mathrm{Lin}(V,W)$ of linear maps $V \to W$ itself forms a linear space, in fact a linear subspace of the linear space of *all* maps $V \to W$. This construction is so important in what follows that we will isolate it for convenient reference: *If* V *and* W *are linear spaces, then the set of all linear maps* $L : V \to W$ *forms a linear space, denoted by* $\mathrm{Lin}(V,W)$.

Kernel and image

For a linear map $L : V \to W$ let the *kernel* of L denote the set of all elements v in V which are taken to the zero element in W.

Linearity of L implies that the kernel $\ker(L)$ is a linear subspace of V, and that L is *injective* precisely when $\ker(L) = \{0\}$ in V. Thus the 'size' of $\ker(L)$ (for example, its dimension: see below) measures the failure of injectivity of L.

The image $\mathrm{im}(L)$ is a linear subspace of W, and by definition L is *surjective* precisely when $\mathrm{im}(L)$ is all of W.

If $L : V \to W$ is a linear map and is at the same time a *bijection* then it is a simple exercise to check that its inverse $L^{-1} : W \to V$ must also be a linear map. (Compare this with the analogous situation for continuous maps between topological spaces; recall §1.5 and Theorem B of §1.6.) A bijective linear map is called a *linear isomorphism*, or just *isomorphism* if the linear context is understood. If $L : V \to W$ is an isomorphism the linear structure of V corresponds precisely to that of W, via L. Thus if there exists an isomorphism between two given linear spaces they are indistinguishable as linear spaces: they are said to be *isomorphic*.

EXAMPLE

Suppose S, T are two linear subspaces of V such that $S + T = V$ and $S \cap T = \{0\}$. Then it turns out that every x in V can be written in a *unique* way as $x = s + t$ where $s \in S$, $t \in T$. We can regard s, t as 'co-ordinates' for x, and it is routine to check that V is isomorphic to the cartesian product $S \times T$. We write $V = S \oplus T$, the *direct sum* of S and T.

Dimension

In euclidean space geometrical problems can be converted into algebraic problems by taking co-ordinates, sometimes already given and sometimes constructed artifically with the solution of a particular problem in mind. This is such a useful technique that it is worth exploiting as much as possible. It leads to the concepts of *basis* and *dimension* in general linear space theory.

A linear space V is said to be *finite-dimensional* if it contains a finite collection $\{e_1, e_2, \ldots, e_m\}$ of elements with the property that each x in V can be written in the form

$$x = \alpha_1 e_1 + \alpha_2 e_2 + \ldots + \alpha_m e_m$$

for some *unique* choice of scalars $\alpha_1, \alpha_2, \ldots, \alpha_m$ (which of course depends on x). The collection $\{e_1, e_2, \ldots, e_m\}$ is called a *basis* for V, and the number m is called the *dimension* of V. It can be shown to be independent of the actual basis considered.

A linear space which is not finite dimensional is said to be *infinite-dimensional*. Note that this says nothing about the existence of any 'infinite basis' for V, whatever that may mean.

EXAMPLES

1. $V = R^n$ has a basis $\{e_1, e_2, \ldots, e_n\}$ where e_i is the vector $(0, 0, \ldots, 1, \ldots, 0)$ with 1 in the ith place and 0 elsewhere. This is called the *standard basis*. Hence the dimension of R^n is n (as we would hope!).

2. Let V be the linear space consisting of all maps $A \to W$ where W is some given linear space, A is an arbitrary set, and addition and scalar multiplication in V are defined pointwise. Then V is finite-dimensional precisely when W is finite-dimensional *and* A is a finite set. For example, when $W = R$ and $A = \{1, 2, \ldots, n\}$ an element of V can be thought of as an n-tuple $(x_1, x_2, \ldots, x_n)$ with each x_i in R,

and V is isomorphic to R^n. If $A = N = \{1, 2, 3, ...\}$ then an element of V is an infinite sequence $(x_1, x_2, x_3, ...)$ of elements of R, and V is infinite-dimensional.

3. Any linear space isomorphic to a finite-dimensional space is itself finite-dimensional, since the isomorphism transports a basis in one space to a basis in the other.

Thus any infinite-dimensional space cannot be isomorphic to a finite-dimensional one. On the other hand, any two finite-dimensional spaces of the *same dimension* are isomorphic. From this and Example 1 it follows that *every finite-dimensional (real) linear space of dimension* n *is isomorphic to* R^n - although in practice the construction of an isomorphism may be rather artificial.

After this excursion into the theory of linear spaces and linear maps we now turn again to the task of generalizing the definition of differentiability.

If V, W are linear spaces and $f : V \to W$ is any map (not necessarily linear) then given x and h in V we can certainly define

$$F(h) = f(x+h) - f(x)$$

as an element of W. Now if we wish to discuss approximating F by some suitable linear map L we need to introduce some metric or at least topological structure. Although we could make a definition of differentiability at x involving only topologies on V and W, experience shows that in the first instance it is useful to ensure that V and W are each *metric spaces* (as well as linear spaces), and moreover that the metrics are in a certain sense compatible with the linear structures. Pursuing these ideas we arrive at the notion of a *normed linear space*.

2.3 NORMED LINEAR SPACES

Observe that in R^n with the euclidean metric the following rules hold:

(i) $d(x+z, y+z) = d(x,y)$ for all x, y, z in R^n (i.e. translation does not alter relative distances).

(ii) $d(\alpha x, \alpha y) = |\alpha| d(x,y)$ for all x, y in R^n and all scalars α (i.e. scalar multiplication magnifies or contracts distances uniformly).

These are not consequences of the general definition of metric (§1.2), since they involve the linear structure of R^n which need not exist in an arbitrary metric space.

Let us say (temporarily) that if V is a linear space with some metric d on it then d is *sensible* if it satisfies (i) and (ii). If d is a sensible metric then from (i) we immediately see that $d(x,y) = d(x-y, 0)$ for every x, y in V. Denoting $d(x,0)$ by $||x||$ (thought of as the length of the 'vector' $\vec{0x}$) we see that in view of (ii) above we have

(1) $||\alpha x|| = |\alpha|\ ||x||$ for every x in V and scalar α,

and then since d is a metric we also have

(2) $||x|| \geq 0$ for all x in V, and $||x|| = 0$ precisely when x is the zero element in V;

(3) $||x+y|| \leq ||x|| + ||y||$ for all x, y in V.

Note that the given d is recoverable from $||\cdot||$ by the formula

$$d(x,y) = ||x-y|| .$$

A function $||\cdot|| : V \to R$ satisfying (1), (2) and (3) is called a *norm* for V, and V together with a particular norm is called a *normed linear space*. Any normed linear space is automatically a metric space, with metric defined by $d(x,y) = ||x-y||$: it is straightforward to check that this d does obey rules (1), (2), (3) of the definition of a metric.

EXAMPLES of normed linear spaces

1. R^n with $||x|| = \left(\sum_{i=1}^{n} |x_i|^2\right)^{\frac{1}{2}}$, i.e. the usual 'length' function for n-vectors. This gives the euclidean metric and so it is called the *euclidean norm*. In particular, when $n = 1$ we have $||x|| = |x|$.

2. (a) R^n with $||x|| = \max\{|x_1|, |x_2|, \ldots, |x_n|\}$,

   (b) R^n with $||x|| = |x_1| + |x_2| + \ldots + |x_n|$.

In fact it can be proved that *any* norm on R^n will give rise to the same topology on R^n as does the euclidean norm. Of course the metric will in general be different.

3. The space of bounded functions $f : R \to R$, with

$$||f|| = \sup\{|f(x)| \mid x \in R\} .$$

This is a linear subspace of the space of *all* functions $R \to R$ (Example 3, §2.2) with pointwise addition and scalar multiplication, and the metric is that of Example 4, §1.2. More generally:

4. The space of bounded maps $f : A \to W$, where W is a normed linear space and A is any set whatsoever. (*Bounded* means there exists a number K such that $||f(x)||_W$ is less than K for all x in A; equivalently, $f(A) \subset B_K(0)$.) Again the obvious norm is

$$||f|| = \sup\{||f(x)||_W \mid x \in A\} .$$

5. The space of continuous functions $f : [0,1] \to R$ with $||f|| = \sup\{|f(x)| \mid a \leq x \leq b\}$. Note that this *is* a linear space with pointwise addition and scalar multiplication as usual, and the norm makes sense in view of Theorem A of §1.6. This space is a linear subspace of the special case of Example 4 above where $A = [0,1]$, and it inherits the norm and metric from this larger space.

6. The space of continuous functions $f : [0,1] \to R$ which are differentiable at each point in $(0,1)$ and whose derivative $f' : (0,1) \to R$ is a bounded function. Here we could take $||f|| = ||f||_0$ where

$$||f||_0 = \sup\{|f(x)| \mid a \leq x \leq b\}$$

or, if we wished to measure derivatives as well as we would take $||f|| = ||f||_1$ where

$$||f||_1 = \sup\{|f(x)| \mid a \leq x \leq b\} + \sup\{|f'(x)| \mid a \leq x \leq b\} \ .$$

This is a linear subspace of the space in Example 5. The norm which it naturally inherits is $||\cdot||_0$, but the norm which it may be most useful to work with is $||\cdot||_1$. The metric corresponding to $||\cdot||_1$ is that of Example 5, §1.2.

7. The space of systems of differential equations

$$\begin{aligned} \dot{x}_1 &= f_1(x_1, x_2) \\ \dot{x}_2 &= f_2(x_1, x_2) \end{aligned} \qquad (F)$$

defined on the closure $\bar{U}$ of a bounded open subset U of R^2 and such that f_1, f_2 are continuous, where the linear structure is defined by pointwise addition and scalar multiplication of maps $f = (f_1, f_2) : R^2 \to R^2$, and the norm is given by

$$||F||_U = \sup\{||(f_1(x), f_2(x))|| \mid x = (x_1, x_2) \in U\} \ .$$

Theorem C of §1.6 shows that $\bar{U}$ is compact, and then Theorem A' guarantees that $||\cdot||_U$ makes sense. It is routine to verify that $||\cdot||_U$ is a norm.

The corresponding metric is like that of Example 6, §1.2, although there we supposed f_1, f_2 to be bounded on the whole of R^2.

Now we have essentially everything we need to pursue the study of differentiability. However, it is worth going a little further at this stage to provide some machinery that will be useful later on.

Linearity and continuity

Every linear map $L : R^n \to R^m$ (with the euclidean norms) is continuous. The proof is easy: let $\{e_1, e_2, \ldots, e_n\}$ be the standard basis for R^n and observe that if x and y are in R^n then

$$
\begin{aligned}
L(x) - L(y) &= L(x-y) && \text{by linearity} \\
&= L\left(\sum_{i=1}^{n} (x_i - y_i)e_i\right) \\
&= \sum_{i=1}^{n} (x_i - y_i)L(e_i) && \text{by linearity.}
\end{aligned}
$$

Hence $||L(x) - L(y)|| \leqslant \sum_{i=1}^{n} |x_i - y_i| \, ||L(e_i)|| \leqslant ||x-y||M$ where

$M = \sum_{i=1}^{n} ||L(e_i)||$. Therefore $||L(x) - L(y)||$ is less than ε whenever $||x-y||$ is less than $\delta = \varepsilon M^{-1}$.

In this proof we have used strongly the fact that the domain R^n is finite-dimensional, but note that the proof would have worked with the codomain replaced by any normed linear space F.

In general a linear map between two normed linear spaces need not be continuous. Here is an example with, necessarily, an infinite-dimensional domain.

EXAMPLE

Let V be the space of all infinite sequences

$$\alpha = (a_1, a_2, a_3, \ldots)$$

with addition and scalar multiplication defined term-wise, and let E be the linear subspace consisting of those with $\sum_{n=1}^{\infty} |a_n|$ finite. Define a norm for E by

$$||\alpha|| = \sum_{n=1}^{\infty} \frac{1}{n!} |a_n| ,$$

and define a linear map $L : E \to E$ by

$$L(a_1, a_2, a_3, \ldots) = (a_2, a_3, a_4, \ldots) .$$

Then L is clearly linear, but is *not* continuous at 0. To see this let $\alpha^{(n)}$ denote the sequence with $(n-1)!$ in the nth place and zero elsewhere. Then $||\alpha^{(n)}|| = \frac{1}{n}$ but $||L(\alpha^{(n)})|| = 1$, and so no matter how small δ is there always exists some $\alpha^{(n)}$ in E with $||\alpha^{(n)}||$ less than δ but $||L(\alpha^{(n)})|| = 1$.

The basic facts about continuity of linear maps are summarized in the following theorem. The details of the proof are straightforward, and can be found in any book (such as Simmons [118]) on functional analysis.

THEOREM

Let $L : E \to F$ *be a linear map between normed linear spaces. Then the following statements are mutually equivalent:*

(1) L *is continuous.*

(2) L *is continuous at the origin in* E.

(3) *There exists a constant* K *such that* $||L(x)||_F$ *is less than* K *for every* x *in* E *with* $||x||_E = 1$.

Statement (3), can be described by saying that however much L distorts the 'unit sphere' S_E in E it nevertheless maps it inside some sphere of radius K and centre the origin in F.

If $L : E \to F$ is a continuous linear map then there is a greatest lower bound for all the constants K which satisfy (3). This is the radius of the smallest sphere centred at the origin in F and containing $L(S_E)$, in other words

$$\sup\{||L(x)||_F \mid ||x||_E = 1\} .$$

This number is denoted by $||L||$, for it can be shown without much trouble that it does define a norm on the linear subspace of Lin(E,F) consisting of *continuous* linear maps.

Note that for such a map L we have the inequality

$$||Lx||_F \leq ||L|| \; ||x||_E$$

for every x in E. This follows immediately from the definition of $||L||$ by writing $||x||_E = \alpha$ and applying L to $\alpha^{-1}x$: we have

$$||L(\alpha^{-1}x)||_F \leq ||L||$$

since $||\alpha^{-1}x||_E = 1$, but the left hand side is $\alpha^{-1}||L(x)||_F$ by the linearity of L and property (1) of the norm.

The notion of the normed linear space of continuous linear maps $E \to F$, with particular properties of the norm such as the inequality above, will be important in §2.6 when we consider higher orders of differentiation. Quite apart from these uses, however, it constitutes one of the basic pieces of equipment in functional analysis.

Completeness

Recall that in a metric space S we say a sequence $x_1, x_2, x_3, \ldots$ converges to c in S when:

> given any $\varepsilon > 0$ there exists a positive integer m such that $d(x_n, c) < \varepsilon$ for all n with $n > m$.

If S is a normed linear space we have $d(x_n, c) = ||x_n - c||$, and if $S = R$ this is just $|x_n - c|$.

When we are given a sequence how can we tell whether or not it converges to some c? If we have a plausible guess for c then we can look at $d(x_n, c)$ and try to check the definition, but what methods are available if we are not sure even whether c exists? In the case $S = R$ there is a very useful lemma (often known as the *General Principle of Convergence*) stating that a sequence of numbers $x_1, x_2, x_3, \ldots$ converges (to something) if and only if

> given any $\varepsilon > 0$ there exists a positive integer m such that $\left|x_{n_1} - x_{n_2}\right| < \varepsilon$ for all $n_1, n_2 > m$.

Unfortunately the generalization of this to arbitrary metric spaces with $d(x_{n_1}, x_{n_2})$ replacing $\left|x_{n_1} - x_{n_2}\right|$ turns out to be false: take, for example, S to be an open interval or to be the set of rational numbers with the metric induced from R. In these cases a sequence as above in S will converge to something in R which may not belong to S.

Metric spaces in which this General Principle of Convergence *does* hold are sufficiently important and common for it to be worthwhile identifying them with a particular name: they are called *complete* metric spaces. A normed linear space which is also complete when viewed as a metric space is known as a *Banach space* (S. Banach, 1892-1945).

The examples given above of non-complete metric spaces are, of course, not (real) linear spaces. In fact it would be impossible to construct a finite-dimensional linear example in view of the result that *every finite-dimensional normed linear space is automatically a Banach space*. Completeness of a normed linear space is a technical condition needed for the correct statement of some general results later on, but in finite dimensions there will be no need to worry about it explicitly.

Hilbert spaces

Apart from its linear and norm structures, R^n has interesting geometrical properties that relate to the *inner product*

$$\langle x,y\rangle = x_1y_1 + x_2y_2 + \dots + x_ny_n$$

of pairs of elements x, y familiar as the *dot product* or *scalar product* for vectors in R^3. We have $\langle x,x\rangle = ||x||^2$, and in R^3 it can be shown that

$$\langle x,y\rangle = ||x|| \; ||y|| \cos\theta$$

where θ is the angle between the vectors x, y. In R^n we can take this as the *definition* of θ, and mimic much of the geometry of R^3.

In functional analysis there are many circumstances under which a Banach space comes provided with an inner product analogous to that in R^n. Technically, an *inner product* on a real normed linear space E means a map $\langle\cdot,\cdot\rangle : E \times E \to R$ which is linear in each factor separately, is symmetric ($\langle x,y \rangle = \langle y,x \rangle$), and is such that $\langle x,x \rangle$ is non-negative for every x, vanishing only when $x = 0$. (In the complex case there are some amendments to these conditions.) Using such an inner product it is possible to derive much stronger results about the structure of a Banach space than would be possible without it. A Banach space which has an inner product is called a *Hilbert space* (D. Hilbert, 1862-1943).

Sometimes it is necessary to ensure that a Hilbert space, although possibly infinite-dimensional, is still not too large in a topological sense. A topological space is called *separable* if it contains a countable dense subset (as with the countable set $\mathbb{Q}$ of rational numbers contained in $\mathbb{R}$). It is particularly useful to deal with separable Hilbert spaces, since it can be proved that they have in a certain precise sense at most countably infinite dimensions, and many finite-dimensional techniques can be extended to this case by replacing n-tuples $(x_1, x_2, \ldots, x_n)$ by infinite sequences.

2.4 DIFFERENTIATION (continued)

Let E and F be two normed linear spaces with respective norms $||\cdot||_E$, $||\cdot||_F$, suppose $f : E \to F$ is a given continuous (not necessarily linear) map, and let x be a particular point of E. By analogy with §2.1 we make the following definition:

DEFINITION

The map f *is* differentiable *at* x *if there is a linear map* $L : E \to F$ *which approximates* f *at* x *in the sense that for all* h *in* E *we have*

$$f(x+h) - f(x) = L(h) + ||h||_E\, \eta(h)$$

where $\eta(h)$ *is an element of* F *with* $||\eta(h)||_F \to 0$ *as* $||h||_E \to 0$.

In general L of course depends on x, and so to be formally correct we should write something like L_x instead of L. Here $||h||_E$ plays the role of $|h|$ in §2.1 while $\eta(h)$ plays the role of ε which is no longer a number but an element (thought of as 'small') of F. It is not difficult to show that if such an $L = L_x$ exists (for given E, F, f and x) then it is unique. We call it the *derivative* of f at x and denote it by Df(x). It is a *linear map* from E to F, and *not a number*. To avoid proliferation of brackets we will often write Df(x).h or Df(x)h instead of Df(x)(h). Remember that h is an element of E, Df(x) is a linear map from E to F, and so Df(x)h is an element of F.

Using this notation we can re-write the defining expression for the derivative as

$$f(x+h) - f(x) = Df(x)h + ||h||\eta(h) \qquad (*)$$

where $||h||$ means $||h||_E$, and $\eta(h) \to 0$ (in F) as $h \to 0$ (in E).

The pictorial interpretation is much as before, although the spaces now may not be 1-dimensional or even finite-dimensional. Instead of thinking of the graph of f (which will be some kind of 'surface' in the cartesian product $E \times F$) it can often be more useful to visualize the image of f. This consists of some distorted version of E inside F. The image of Df(x) is a linear subspace of F, and when translated by the vector f(x) it is 'tangent' to the image of f at the point f(x). See Figure 9.

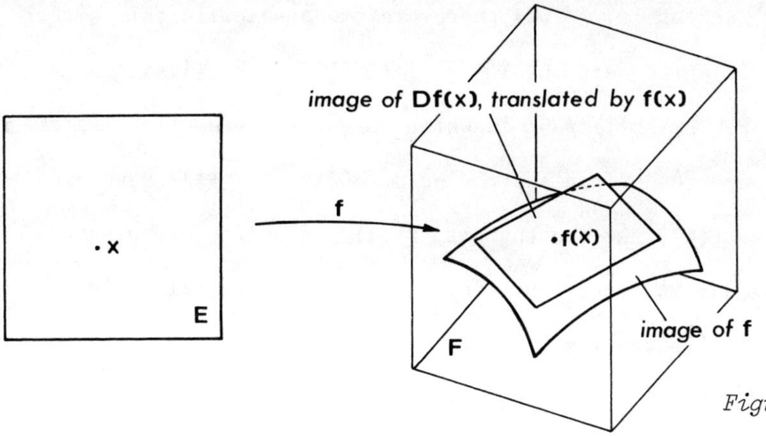

Figure 9

Equivalently, the map

$$h \mapsto f(x) + Df(x)h$$

is a *first order approximation* to the map $h \mapsto f(x+h)$.

Interpretation of the derivatives in familiar cases

We will now look at the meaning of the above rather abstract definition in the case of maps between euclidean spaces, where differentiation is a more familiar concept.

Case 1. $E = R$, $F = R^m$.

Here f is a continuous map $R \to R^m$, which we call a *path* in R^m . (From some points of view it would be more sensible to give that name to the *image* of f in R^m , but then we would have lost track of its parametrization by R. We shall continue to call the map a path.)

At each point t in R the derivative $Df(t)$, if it exists, is a linear map $R \to R^m$. Now linear maps L from R to any linear space F are very easy to understand, since for any s in R we have $s = s.1$ and so by linearity of L we have $L(s) = sL(1)$. In other words, to find $L(s)$ all you have to do is to find $L(1)$ as a element of F, and multiply it by the scalar s. Thus there are two possibilities: either $L(1) = 0$ in F, in which case $L(s) = 0$ for all $s \in R$ (i.e. L is the *zero map* $R \to F$) or $L(1) \neq 0$ in which case the image of L is the straight line through the origin and the point $L(1)$ in F. Returning to the case when $L = Df(t)$, we see that if $Df(t)$ is not the zero map $R \to R^m$ then its image is a straight line which, after the origin is translated to $f(t)$, is tangent at $f(t)$ to the image of the path f. See Figure 10.

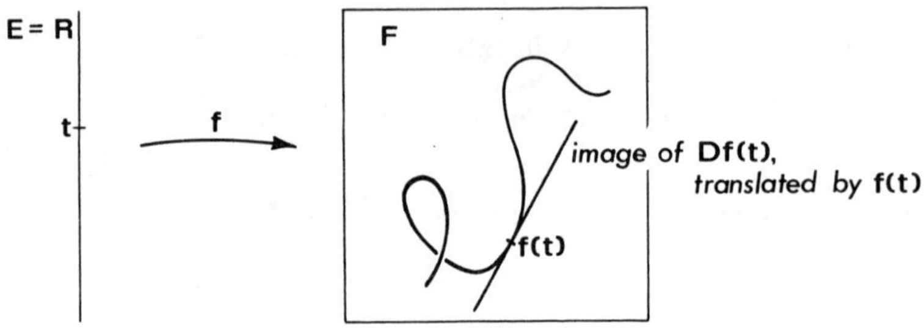

Figure 10

The association of $L(1)$ with L sets up a 1-1 correspondence between *linear maps* $R \to F$ and *elements of* F, and it is a simple matter to check that this is a linear isomorphism between the space of all linear maps $R \to F$ (pointwise addition and scalar multiplication) and the space F itself. For this reason whenever we have a map $f : R \to F$ (in particular when $F = R^n$) it is often convenient to work not with $Df(t)$ but $Df(t)(1)$ or $Df(t)1$, which is an element of F and denoted by $f'(t)$. By linearity we have for all s in R

$$Df(t)s = Df(t)(s.1) = sDf(t)1 = sf'(t)$$

and so the formula (*) defining the derivative of $f : R \to F$ becomes

$$f(t+h) - f(t) = hf'(t) + ||h||\eta(h)$$

where $\eta(h) \to 0$ as $h \to 0$. Finally, returning to the specific case when $F = R^m$ we have $f(t) = (f_1(t), f_2(t), \ldots, f_n(t))$ and the above formula shows us that $f'(t)$ is simply the n-vector $(f_1'(t), f_2'(t), \ldots, f_n'(t))$ where the f_i' are the usual derivatives of the real-valued functions f_i. This fits neatly into Figure 10, illustrating our familiar notion of $(f_1'(t), f_2'(t), \ldots, f_n'(t))$ as a vector based at $f(t)$ and tangent to the (image of the) path there. Of course, Figure 10 serves equally well as a pictorial representation of the case for general F, not necessarily R^m.

To round off this description we look at the case $n = 1$, bringing us back to the beginning of the discussion of differentiation (§2.1). We have two concepts of derivative: the old familiar $f'(t)$ as a number, and the new $Df(t)$ as a linear map $R \to R$. What is their relationship? From the above we see that it is merely that

$$f'(t) = Df(t)1 \quad ,$$

the right hand side being a number (element of R) since $n = 1$.

Case 2. $E = R^n$, $F = R$.

Here f is a real-valued function of n real variables $(x_1, x_2, \ldots, x_n) = x$, and $Df(x)$ (if it exists) is a linear map $R^n \to R$. To analyze this linear map we will take the standard basis $\{e_1, e_2, \ldots, e_n\}$ for R^n (see §2.2). Then any h in R^n can be written uniquely as

$$h = h_1e_1 + h_2e_2 + \ldots + h_ne_n$$

(i.e. $h = (h_1, h_2, \ldots, h_n)$) and so by linearity we have

$$\begin{aligned} Df(x)h &= h_1Df(x)e_1 + h_2Df(x)e_2 + \ldots + h_nDf(x)e_n \\ &= h_1\alpha_1 + h_2\alpha_2 + \ldots + h_n\alpha_n \text{ , say.} \end{aligned}$$

By taking h to be of the form $(0,0,\dots,h_i,\dots,0)$ (i.e. h_i in the ith place and zeros elsewhere) for each i in turn, and substituting in the formula (*), we find that the α_i are nothing other than the familiar *partial derivatives* $\frac{\partial f}{\partial x_i}$ at the point x.

If v is any element of R^n then the number $Df(x)v$ is called the (partial) derivative of f at x in the direction of v. Thus in this terminology the partial derivative $\frac{\partial f}{\partial x_i}(x)$ is the derivative of f at x in the direction of e_i. If R^n is replaced by an arbitrary normed linear space E then the notion of 'standard' partial derivatives $\frac{\partial f}{\partial x_i}$ will not exist, but the derivative of f in the direction of v (where v is an element of E) will still make sense.

Case 3. $E = \mathsf{R}^n$, $F = \mathsf{R}^m$.

In this case $Df(x)$ is a linear map $\mathsf{R}^n \to \mathsf{R}^m$, and so if we choose bases $\{e_1,e_2,\dots,e_n\}$, $\{\varepsilon_1,\varepsilon_2,\dots,\varepsilon_m\}$ (not necessarily the standard bases) for R^n and R^m respectively we can represent $Df(x)$ by a matrix. Specifically, if

$$h = \ell_1 e_1 + \ell_2 e_2 + \dots + \ell_n e_n$$

then

$$Df(x)h = \lambda_1 \varepsilon_1 + \lambda_2 \varepsilon_2 + \dots + \lambda_m \varepsilon_m$$

where the λ's are related to the ℓ's by

$$\lambda_i = \sum_{j=1}^{n} \alpha_{ij}\, \ell_j\,, \qquad 1 \leqslant i \leqslant m$$

for a collection of mn scalars $\{\alpha_{ij}\}$ forming a matrix with m rows and n columns.

If we now do take the bases to be the standard bases, so that the ℓ_i and λ_i are the usual coordinates, then writing

$$f(x) = (f_1(x), f_2(x), \ldots, f_m(x))$$

and putting together the discussions of cases 1 and 2 we find that

$$\alpha_{ij} = \frac{\partial f_i}{\partial x_j}(x)$$

for each i, j. The matrix representing $Df(x)$ with respect to the standard bases is thus

$$\begin{pmatrix} \frac{\partial f_1}{\partial x_1}(x) & \frac{\partial f_1}{\partial x_2}(x) & \cdots & \frac{\partial f_1}{\partial x_n}(x) \\ \frac{\partial f_2}{\partial x_1}(x) & \frac{\partial f_2}{\partial x_2}(x) & \cdots & \frac{\partial f_2}{\partial x_n}(x) \\ \vdots & \vdots & & \vdots \\ \frac{\partial f_m}{\partial x_1}(x) & \frac{\partial f_m}{\partial x_2}(x) & \cdots & \frac{\partial f_m}{\partial x_n}(x) \end{pmatrix}.$$

This is the so-called *Jacobian matrix* (with respect to standard bases) condensed to the notation

$$\frac{\partial(f_1, f_2, \ldots, f_m)}{\partial(x_1, x_2, \ldots, x_n)}.$$

Observe that we could equally well construct an analogous Jacobian matrix for other choices of bases.

When $m = n$ the matrix is a square matrix, and so has a determinant. This number is often denoted by $Jf(x)$ or just J when f and x are understood: it will be of considerable interest to us later, in §2.8.

Remarks

1. The Jacobian matrix depends on a choice of bases, whereas the derivative $Df(x)$ does not. The derivative was defined entirely without reference to coordinates.

2. Although the differentiability of f at x implies the existence of the partial derivatives of the components f_i of f, it is a standard fact that the converse does not hold. This is essentially because the partial derivatives tell us approximately how the function behaves in n particular directions, whereas the derivative $Df(x)$ deals with the behaviour in a whole neighbourhood. It can be proved, however, that if the partial derivatives exist and are themselves continuous functions in some neighbourhood of x then f is differentiable at x. We shall not be concerned with technicalities of this kind in what follows.

3. *Digression into the complex field.* All the above ideas work just as well for maps between *complex* linear spaces. In particular, if C^n denotes the linear space of n-tuples of complex numbers with co-ordinate wise addition and multiplication by scalars (complex numbers) and norm given by

$$||z|| = \left(\sum_{i=1}^{n} |z_i|^2\right)^{\frac{1}{2}}$$

then the derivative of a differentiable map $f : C^n \to C^m$ at z will be a complex-linear map $Df(z) : C^n \to C^m$. Now C^n can be thought of as R^{2n} by, for example, identifying $(z_1, z_2, \ldots, z_n)$ with $(x_1, y_1, x_2, y_2, \ldots, x_n, y_n)$ where $z_j = x_j + iy_j$ for each $j = 1, 2, \ldots, n$, although in C^n we can multiply by complex scalars while in R^{2n} we can multiply only by real scalars. A complex-linear map $C^n \to C^m$ becomes a real linear map $R^{2n} \to R^{2m}$ (since real scalars are special cases of complex scalars) but the converse is not true. At z in C^n the conditions for a real-differentiable map $f : R^{2n} \to R^{2m}$ to represent a complex-differentiable map $f : C^n \to C^m$ are that

$$\frac{\partial u_i}{\partial x_j} = \frac{\partial v_i}{\partial y_j}, \quad \frac{\partial u_i}{\partial y_j} = -\frac{\partial v_i}{\partial x_j} \qquad \begin{matrix} 1 \leqslant i \leqslant m \\ 1 \leqslant j \leqslant n \end{matrix} \qquad \text{at } z,$$

where we write $f(x_1,y_1,x_2,y_2,\ldots,x_n,y_n) = (u_1,v_1,u_2,v_2,\ldots,u_m,v_m)$, each u_k,v_k being a function of the x_i and y_j. These are the classical *Cauchy-Riemann equations*.

So far we have been discussing maps f whose domain E and codomain F are both normed linear spaces, but it is clear that for differentiability to be defined at x it is not necessary that f be defined on the whole of E. All we need is that $f(x+h)$ should be defined for all sufficiently small h, i.e. on some δ-ball $B_\delta(x)$. Hence if the domain of f is given as an *open set* U in a normed linear space E it makes sense to talk about differentiability of f at any point x in U. Note, however, that although the formula (*) only has meaning if h is such that $x + h$ lies in U, the map $Df(x)$ is linear and so its domain can be thought of as the whole of E. Thus associated to

$$f : U \to F$$

we have

$$Df(x) : E \to F \qquad \text{(linear)}$$

for every x in U at which f is differentiable.

If it is the case that f is differentiable at x for *every* x in U then we simply say $f : U \to F$ is *differentiable*, or that f is *differentiable on* U. For each x in U we have a linear map $Df(x) : E \to F$, or in other words we have a map

$$Df : U \to \{\text{linear maps } E \to F\} = \text{Lin}(E,F) \quad .$$

At this point we may wish to ask whether the derivative of f at x varies continuously with x, i.e. whether the map $Df : U \to \text{Lin}(E,F)$ is continuous. To make sense of this we would have to have a topology on $\text{Lin}(E,F)$, or at least on some subset of $\text{Lin}(E,F)$ containing the image of Df. Now it so happens that if the continuous map $f : U \to F$ is differentiable at x then $Df(x)$ is not merely a linear map $E \to F$ but is a *continuous* linear map (i.e. continuous on E, for each fixed x). This means that the image of $Df : U \to \text{Lin}(E,F)$ is contained in the subspace of $\text{Lin}(E,F)$ consisting of continuous linear maps. As we saw in §2.3 this space, which we will now denote by $L(E,F)$, has a natural structure as a normed linear space and hence a metric space. Therefore if we regard Df not just as a map

$$Df : U \to \text{Lin}(E,F)$$

but as a map

$$Df : U \to L(E,F)$$

we are able to discuss whether or not it is continuous.

DEFINITION

If Df *is continuous we say that* $f : U \to F$ *is of class* C^1, *or* f *is of class* C^1 *on* U *- or simply that* f *is* C^1 .

This somewhat abstract formulation may seem far removed from familiar ideas of continuity of derivative in euclidean space, but the gap is easily bridged. To begin with, recall that there is no distinction between $\text{Lin}(R^n,R^m)$ and $L(R^n,R^m)$ since, as we have seen, every linear map $R^n \to R^m$ is continuous. There are various reasonable ways to put a norm on $L(R^n,R^m)$. First of all we can take the norm defined above in the general context, namely

$$||L|| = \{\sup||L(x)|| \;\Big|\; ||x|| = 1\}$$

where the norms on the right hand side refer to the usual euclidean norms. Secondly, we could choose norms other than euclidean norms for R^n, R^m and make the analogous definition of $||L||$. Thirdly, we could choose bases for R^n and R^m, then represent L by an $m \times n$ matrix $A = (a_{ij})$ and, regarding A as an element of R^{mn}, take $||L||$ to be the euclidean norm of A, i.e.

$$||L|| = ||A|| = \left(\sum_{i=1}^{m} \sum_{j=1}^{n} |a_{ij}|^2\right)^{\frac{1}{2}} .$$

Then there are other possible variants such as

$$||L|| = \max_{i,j} |a_{ij}|$$

and so on. Now the important fact to use here, which sweeps away all this confusion, is that *it makes no difference which norm we choose on* $L(R^n,R^m)$. More precisely, since $L(R^n,R^m)$ is finite-dimensional we can invoke a non-trivial theorem (Simmons [118]) which says that *any two norms on a finite-dimensional linear space will induce the same topology*, although of course the induced *metrics* will in general be different.

Consequently, for studying the continuity (a purely topological property) of $Df : U \to L(R^n,R^m)$ we can choose whichever way we like of measuring $||Df(x)||$. A convenient way is to take standard bases and let

$$||Df(x)|| = \max_{i,j} \left|\frac{\partial f_i}{\partial x_j}(x)\right| .$$

It is then simple to see that continuity of Df is the same as continuity of the partial derivatives $\frac{\partial f_i}{\partial x_j}$. Therefore $f : U \to R^m$ (U an open subset of $\mathbb{R}^n$) is C^1 precisely when *all the* $\frac{\partial f_i}{\partial x_j}$ *are continuous functions on* U.

This freedom in the choice of norm is another illustration of the advantages of working with topologies rather than particular metric structures.

2.5 PROPERTIES AND USES OF THE DERIVATIVE

Unless stated otherwise, E and F will denote normed linear spaces with U an open subset of E.

The following properties of the derivative can be verified directly from the definition:

(i) *Linear combinations*. If $f,g : U \to F$ are differentiable then so is the map $\alpha f + \beta g$ for any constants α, β, and

$$D(\alpha f + \beta g) = \alpha Df + \beta Dg$$

as maps $U \to L(E,F)$. Constant maps themselves have derivative zero, of course.

(ii) *Linear maps*. If f is the restriction to U of a continuous linear map $L : E \to F$ then $Df(x) = L$ for every x in U, i.e. L is already its own linear approximation. In particular f is C^1. (Note that we do *not* say $Df(x) = L(x)$, which would not make sense.)

In the special case $E = F = \mathbb{R}$ we have $f(x) = lx$ for some number l, and $Df(x) = l$ regarded as the linear map $\mathbb{R} \to \mathbb{R} : x \mapsto lx$.

(iii) *Bilinear maps*. If $E = E_1 \times E_2$ and f is the restriction to U of a continuous bilinear map $B : E_1 \times E_2 \to F$ (i.e. B is linear in each factor separately) then

$$Df(x)h = B(x_1,h_2) + B(h_1,x_2)$$

where we write $x = (x_1,x_2)$, $h = (h_1,h_2)$ with x_1,h_1 in E_1 and x_2,h_2 in E_2. In particular f is C^1, since the map $Df : U \to L(E,F)$ taking x to the linear map

$$(h \mapsto B(x_1,h_2) + B(h_1,x_2))$$

is continuous.

As an example, consider $E_1 = E_2 = F = \mathbb{R}$ and $f(x_1,x_2) = x_1x_2$ (ordinary multiplication). Then

$$\begin{aligned} Df(x)h &= x_1h_2 + h_1x_2 \\ &= (x_2,x_1)\begin{pmatrix} h_1 \\ h_2 \end{pmatrix} \end{aligned}$$

which is another way of saying that $\dfrac{\partial f}{\partial x_1} = x_2, \dfrac{\partial f}{\partial x_2} = x_1$.

(iv) *Cartesian products.* If $f_1 : U_1 \to F_1$ and $f_2 : U_2 \to F_2$ are differentiable then so is $f = f_1 \times f_2 : U_1 \times U_2 \to F_1 \times F_2$ where $(f_1 \times f_2)(x_1,x_2)$ means $(f_1(x_1), f_2(x_2))$, and we have

$$Df(x_1,x_2)(h_1,h_2) = (Df_1(x_1)h_1, Df_2(x_2)h_2) .$$

In other words, derivatives operate co-ordinate-wise.

(v) *Compositions.* (Important) Suppose E,F and G are three normed linear spaces, and U,V are open sets in E,F respectively. Let $f : U \to F$ and $g : V \to G$ be continuous maps, and suppose the image of f lies in V so that the composition $g \circ f : U \to G$ exists: see Figure 11.

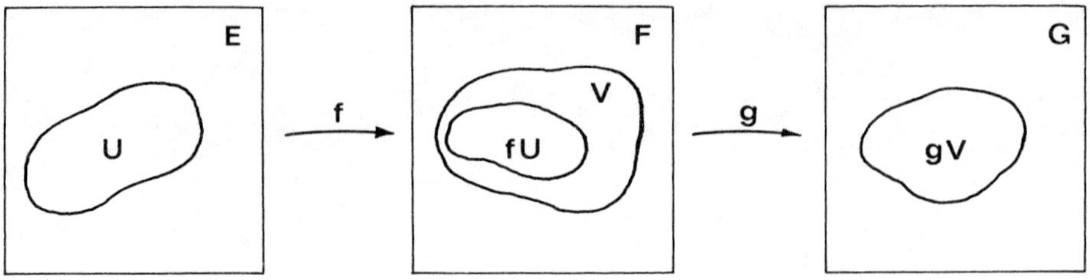

Figure 11

If f is differentiable at x and g is differentiable at $f(x)$ then $g \cdot f$ is differentiable at x and

$$D(g \cdot f)(x) = Dg(f(x)) \cdot Df(x) \quad .$$

Observe that the right hand side is the composition of the two continuous linear maps $Df(x) : E \to F$ and $Dg(f(x)) : F \to G$ and is itself a continuous linear map from E to G.

This formula for the derivative of $g \cdot f$ is known as the Chain Rule. It can be expressed by saying that *the derivative of a composition is the composition of the derivatives*, but note that care must be taken to specify at which points the derivatives are calculated.

In euclidean space the derivatives can be represented by corresponding Jacobian matrices, and then we obtain the familiar chain rule for partial derivatives. Explicitly, if $E = \mathbb{R}^n$, $F = \mathbb{R}^m$, $G = \mathbb{R}^p$ and we write $y = f(x)$, $z = g(y)$, then the chain rule states that

$$\frac{\partial(z_1, z_2, \ldots, z_p)}{\partial(x_1, x_2, \ldots, x_n)} = \frac{\partial(z_1, z_2, \ldots, z_p)}{\partial(y_1, y_2, \ldots, y_m)} \cdot \frac{\partial(y_1, y_2, \ldots, y_m)}{\partial(x_1, x_2, \ldots, x_n)}$$

or equivalently

$$\frac{\partial z_i}{\partial x_j} = \sum_{k=1}^{m} \frac{\partial z_i}{\partial y_k} \cdot \frac{\partial y_k}{\partial x_j}$$

which is the form in which the rule is most commonly expressed. This form is the one which is most useful for specific calculations, but its disadvantages are (1) it is cumbersome, (2) it depends on choices of co-ordinates in R^n etc., (3) it gives (as written) no information about where to evaluate the partial derivatives.

EXAMPLES

1. Ordinary multiplication of real numbers is a bilinear map $m : \mathsf{R} \times \mathsf{R} \to \mathsf{R}$, and the Chain Rule allows us to use this to derive a rule for the derivative of the pointwise product $fg : E \times F \to \mathsf{R}$ of two functions $f : E \to \mathsf{R}$ and $g : F \to \mathsf{R}$. Expressing $f(x)g(y)$ as $m(f(x),g(y))$ we can write $fg = m\circ(f \times g)$ and so by the Chain Rule

$$\begin{aligned} D(fg)(x,y) &= Dm(f(x),g(y)) \circ D(f \times g)(x,y) \\ &= Dm(f(x),g(y)) \circ (Df(x) \times Dg(y)) \end{aligned}$$

by the rule for cartesian products. Applying both sides to (h,k) in $E \times F$ we obtain

$$\begin{aligned} D(fg)(x,y).(h,k) &= Dm(f(x),g(y))\circ (Df(x)h,Dg(y)k) \\ &= m(f(x),Dg(y)k) + m(Df(x)h,g(y)) \end{aligned}$$

by the rule for bilinear maps. The right hand side is simply $f(x)Dg(y)k + g(y)Df(x)h$ (multiplication is commutative), and so we can write the conclusion as

$$D(fg)(x,y) = f(x)Dg(y) + g(y)Df(x)$$

where here the right hand side is a pointwise linear combination of the two linear maps $Df(x) : E \to \mathsf{R}$ and $Dg(y) : F \to \mathsf{R}$. In the particular case when $E = F = \mathsf{R}$ this of course reduces to nothing more than the usual product rule

$$(fg)' = fg' + gf' \quad .$$

2. If $B : E \times E \to \mathbb{R}$ is a bilinear function let $f : E \to \mathbb{R}$ be defined by $f(x) = B(x,x)$. Such a function is often called a *quadratic form*. We can write f as the composition $f = B \cdot d$ where d is the 'diagonal' linear map $d : E \to E \times E$ taking x to (x, x). Since $Dd(x) = d$ we have

$$\begin{aligned} Df(x)h &= DB(d(x)) \cdot Dd(x)h \\ &= DB(x,x).(h,h) \\ &= B(x,h) + B(h,x) \quad . \end{aligned}$$

If B is symmetric in the two factors this becomes

$$Df(x)h = 2B(x,h)$$

or $Df(x) = 2B(x,\cdot)$. A particularly common example is the case $E = \mathbb{R}^n$ and $B =$ scalar product , i.e.

$$B(x,y) = x.y = \sum_{i=1}^{n} x_i y_i \quad :$$

here $f(x) = ||x||^2$ and $Df(x)$ is the linear map $h \mapsto 2x.h$. This we can represent by the $1 \times n$ matrix or row-vector $2x$. More generally, if $f(x)$ is the quadratic form $x^T Ax$ then $Df(x)$ is the row-vector $x^T(A+A^T)$ (where ^{T} denotes transpose of the vector x and of the matrix A).

3. A superficially more complicated but in fact equally straightforward class of examples is illustrated by the following: Let A,b be an $m \times n$ matrix and an n-vector, respectively, each varying differentiably with respect to some real parameter t, and let $f(t) = A(t)b(t)$ in $\mathbb{R}^m$. What is the formula for $f'(t)$, or, more precisely $Df(t).1$, in $\mathbb{R}^m$? Writing $f = B \cdot (A \times b)$ where B is the bilinear map

$$B : L(\mathbb{R}^n,\mathbb{R}^m) \times \mathbb{R}^n \to \mathbb{R}^m : (A,b) \mapsto Ab \ ,$$

we have

$$f'(t) = Df(t).1 = DB(A(t),b(t)) \cdot (A'(t),b'(t))$$
$$= B(A(t),b'(t)) + B(A'(t),b(t))$$
$$= A(t)b'(t) + A'(t)b(t)$$

which makes sense because $b'(t) = Db(t).1$ in R^n and $A'(t) = DA(t).1$ in $L(R^n,R^m)$. There are many similar examples that could be constructed using products of matrices, transposes and so on.

(vi) If $f : U \to V$ is a differentiable map which has a differentiable inverse $g : V \to U$ then at each point x in U the derivative $Df(x) : E \to F$ is a linear *isomorphism*. This is easy to prove: we have

$$g \cdot f = id : U \to U$$

and so the Chain Rule gives

$$Dg(f(x)) \cdot Df(x) = D(id)(x)$$
$$= id : E \to E$$

since $id : U \to U$ is (the restriction of) a linear map. Similarly

$$f \cdot g = id : V \to V$$

and so by the Chain Rule at the point $f(x)$ in V

$$Df(x) \cdot Dg(f(x)) = id : F \to F$$

since $Df(g(f(x))) = Df(x)$. Therefore $Df(x)$ has $Dg(f(x))$ as its inverse, and so in particular $Df(x)$ is a linear isomorphism. Again, this can be casually expressed by saying *the derivative of the inverse is the inverse of the derivative.*

A differentiable map with differentiable inverse is called a *diffeomorphism*. A diffeomorphism defined on some open neighbourhood U of a given point x is called a *local diffeomorphism* at x, although this term is often reserved to apply to local diffeomorphisms whose domain and codomain are both open neighbourhoods of x in E and which keep the point x itself fixed. Local diffeomorphisms at x can be regarded simply as invertible (non-linear) changes of local co-ordinates near x.

(vii) A much-used result in elementary calculus is the *Mean Value Theorem*. This states, roughly, that if a line segment is drawn joining two points on the graph of a differentiable function then there is somewhere on the graph in between these points where the slope of the curve is the same as the slope of the line segment. More accurately, if the interval $[x, x+h]$ lies in the domain of the differentiable function f (although strictly speaking we do not need differentiability at the end-points: continuity there suffices) then there exists some point ξ with $x < \xi < x+h$ such that

$$f'(\xi) = (f(x+h) - f(x))/h \quad .$$

Rewriting this as

$$f(x+h) = f(x) + f'(\xi)h$$

gives another interpretation of the theorem, namely that $f(x+h)$ can be expressed exactly as $f(x)$ plus a linear map applied to h. The linear map is the derivative of f at ξ, not that at x which, from the definition, may give only an approximation to $f(x+h)$. This latter interpretation, rather than the 'graph' version, generalizes directly to n dimensions or to any normed linear space:

THEOREM (Mean Value Theorem)

Let $f : U \to R$ *be a differentiable function. Let* x *and* $x+h$ *lie in* U, *and suppose the whole line-segment joining* x *to* $x+h$ *also lies in* U. *Then there is some point* γ *on the line-segment such that*

$$f(x+h) = f(x) + Df(\gamma)h \quad .$$

The proof is easily obtained from the one-dimensional version above. Given x and h in E, let $\phi : R \to R$ be the function defined by

$$\phi(t) = f(x + th)$$

i.e. f evaluated at the point a proportion t along the line segment from x to $x+h$. Then ϕ is differentiable, being the composition of the differentiable functions $R \to E : t \mapsto x + th$ and $f : E \to R$, and the Chain Rule gives

$$\phi'(t) = Df(x + th)h \quad .$$

By the standard Mean Value Theorem the left hand side equals $\phi(1) - \phi(0)$ for t equal to some ξ between 0 and 1. But

$$\phi(1) - \phi(0) = f(x+h) - f(x) \quad ,$$

and so taking $\gamma = x + \xi h$ finishes the proof of the theorem.

The main use of the theorem is as a tool for estimating the difference between $f(x+h)$ and $f(x)$, given h. If we happen to know that $||Df(x)|| < M$ for some number M and *all* x in U, then we deduce from the Mean Value Theorem that

$$|f(x+h) - f(x)| < M||h||$$

since

$$|Df(\gamma)h| \leqslant ||Df(\gamma)|| \; ||h||$$

as we saw on page 50.

Although we shall not be making such estimates explicitly, we shall meet several instances of the Mean Value Theorem being employed in this way.

Finally, we remark that the above properties (i), (iv) and (v) concerning combinations, products and compositions for differentiable maps are equally valid when 'differentiable' is replaced by C^1 .

2.6 HIGHER ORDERS OF DIFFERENTIATION

Recall that to say that $f : U \to F$ is of class C^1 is to say that the map

$$Df : U \to L(E,F)$$

is continuous, where $L(E,F)$ has the topology which comes from the norm

$$||L|| = \{\sup ||L(x)|| \; \Big| \; ||x|| = 1\} \quad .$$

(Note: Do not confuse the possible continuity of Df with the necessary (if f is continuous) continuity of $Df(x) : E \to F$.) We can next ask whether Df itself is differentiable at a given point x in U. If it is, then $D(Df)(x)$ will be a linear map

$$E \to L(E,F) \quad .$$

This is called the *second derivative* of f at x, abbreviated to $D^2f(x)$. Just as the continuity of f implied that its derivative was a continuous linear map, so the fact that Df is continuous implies that $D^2f(x)$ is a continuous linear map, i.e. $D^2f(x)$ belongs to $L(E,L(E,F))$. If Df is differentiable on the whole of U we thus have a map

$$D^2f : U \to L(E,L(E,F)) \quad .$$

If *this* is continuous, which is the same thing as saying that Df is C^1, we say that f is of class C^2. Although the right hand side looks complicated it is still a normed linear space, and we can therefore continue in this way to define C^3, C^4 etc. If f is of class C^r for all $r = 1, 2, 3, \ldots$ then we say that f is of class C^∞. Clearly, if f is C^r then $D^{r-1}f$ exists and is continuous, or we would not have considered differentiating it: thus C^r implies C^{r-1}, and hence C^r implies C^s for all $s < r$. Conventionally we write C^0 to mean *continuous*.

Sometimes the words *smooth* or just *differentiable* are used in the literature to mean C^∞. Later we shall tend to use *smooth* to mean either C^∞ or C^r for some r understood from the context.

Discussion of derivatives of high order rapidly becomes very uncomfortable because we have to deal explicitly with spaces of type

$$L(E,L(E,L(E,\ldots,L(E,L(E,F)))\ldots)\quad .$$

Fortunately we can avoid the problem by making use of some linear isomorphisms which conveniently exist between these spaces and a family of other types of space which are conceptually easier to grasp. These are spaces of *multilinear maps*.

We have already encountered examples of bilinear maps. An n-*multilinear map* is a map

$$M : E_1 \times E_2 \times \ldots \times E_n \to F \quad ,$$

where all the E_i and F are linear spaces, which is a linear map in each factor separately (i.e. when all the components in the other factors are held fixed). For example, the multiplication map $(x_1,x_2,\ldots,x_n) \mapsto x_1x_2\ldots x_n$ is an n-multilinear map $R^n = R \times R \times \ldots \times R \to R$. As another example, consider the space of all $n \times n$ matrices regarded as

$$R^n \times R^n \times \ldots \times R^n$$

n times, each factor corresponding to one column vector. Then the *determinant* map is an n-multilinear map

$$R^n \times R^n \times \ldots \times R^n \to R \quad .$$

As special cases we note that a linear map is a 1-multilinear map and a bilinear map is a 2-multilinear map. Observe that for $n > 1$ an n-multilinear map is in general *not linear*, since the effect of multiplying something on the left hand side by a scalar α is to multiply in F by α^n.

Consider now the space $L(E,L(E,F))$, or more generally $L(E_1,L(E_2,F))$. An element L of this space is a map associating to every x_1 in E_1 an element $L(x_1)$ in $L(E_2,F)$. Now to every x_2 in E_2 the map $L(x_1)$ associates an element $L(x_1).x_2$ of F. Writing $\tilde{L}(x_1,x_2)$ instead of $L(x_1).x_2$ we have constructed a map

$$\tilde{L} : E_1 \times E_2 \to F \quad .$$

Since L and $L(x_1)$ were linear this map $\tilde{L}$ is easily seen to be bilinear. Furthermore, since $L(x_1)$ and L are continuous it can be shown that $\tilde{L}$ is continuous, where $E_1 \times E_2$ is equipped with the product topology (see page 15).

Conversely, if we start with a bilinear map $\tilde{L} : E_1 \times E_2 \to F$ then we can define a corresponding linear map

$$L : E_1 \to \mathrm{Lin}(E_2,F)$$

by taking $L(x_1)$ to be, for each x_1 in E_1, the linear map $E_2 \to F$ taking x_2 to $\tilde{L}(x_1,x_2)$. Now since we have assumed nothing about continuity of $\tilde{L}$ to begin with, there is no reason to expect that $L(x_1)$ will be continuous, or, if it is, that $L : E_1 \to L(E_2,F)$ will be continuous. However, if we do assume that $\tilde{L}$ is continuous then it turns out that $L(x_1)$ and L are continuous and so L belongs to $L(E_1,L(E_2,F))$. Therefore we see that $L \longleftrightarrow \tilde{L}$ gives a natural bijection between $L(E_1,L(E_2,F))$ and the space $M_2(E_1 \times E_2;F)$ of *continuous* bilinear maps $E_1 \times E_2 \to F$, and moreover it is routine to verify that when M_2 is made into a linear space by pointwise addition and scalar multiplication this bijection is a linear isomorphism.

In a similar way, the space

$$L(E_1,L(E_2,L(E_3,\dots,L(E_{n-1},L(E_n,F)))\dots)$$

can be seen to be linearly isomorphic to the space

$$M_n(E_1 \times E_2 \times \dots \times E_n,F)$$

of *continuous* n-multilinear maps with pointwise addition and scalar multiplication: the correspondence is $L \longleftrightarrow \tilde{L}$ where

$$\tilde{L}(x_1,x_2,\dots,x_n) = (\dots((L(x_1))(x_2))\dots(x_{n-1}))x_n \ .$$

This now gives us a much less clumsy notation for dealing with derivatives. Instead of

$$(\dots((D^r f(x)(h_1))(h_2))\dots(h_{r-1}))h_r$$

we simply write

$$D^r f(x)(h_1,h_2,\dots,h_r) \quad ,$$

thinking of $D^r f$ here as a map $U \to M_r(E \times E \times \dots \times E;F)$. When $E = \mathbb{R}^n$ and $F = \mathbb{R}$, so that we are dealing with a real-valued function of n variables, the expression for $D^r f(x)(h,h,\dots,h)$ becomes

$$\sum a_{ij\dots p}(x)\ h_i h_j \dots h_p$$

where

$$a_{ij\dots p} = \frac{\partial^r f}{\partial x_i \partial x_j \dots \partial x_p}$$

and the summation is over all r-tuples (with possible repeats) of the integers from 1 to n. When $n = 1$ this reduces to

$$D^r f(x)(h,h,\dots,h) = f^{(r)}(x)\ h^r$$

where $f^{(r)}$ is the usual rth derivative of a function of one variable. Formally, we see that the relationship between $f^{(r)}(x)$ as a number and the abstract interpretation of $D^r f(x)$ as a multilinear map is that

$$f^{(r)}(x) = D^r f(x)(1,1,\ldots,1) .$$

In other words, we need only one coefficient to describe the effect of an r-multilinear map $M : \mathbb{R} \times \mathbb{R} \times \ldots \times \mathbb{R} \to \mathbb{R}$ on $(h,h,\ldots,h)$, and when $M = D^r f(x)$ this coefficient is $f^{(r)}(x)$.

We remark here that, as with linear maps, every multilinear map is automatically continuous when we work with finite-dimensional spaces. Therefore explicit references to the continuity of $D^r f(x)$ as a multilinear map can be conveniently dispensed with when f is a map between euclidean spaces.

The particular case $r = 2$. Here $D^2 f(x)$ is thought of as a bilinear map $E \times E \to F$. When $E = \mathbb{R}^n$ and $F = \mathbb{R}$ then $D^2 f(x)(\cdot,\cdot)$ is a quadratic form on $\mathbb{R}^n$. Given a choice of basis in $\mathbb{R}^n$, any such bilinear map $B : \mathbb{R}^n \times \mathbb{R}^n \to \mathbb{R}$ can be represented by a matrix $Q = (q_{ij})$, where

$$\begin{aligned} B(u,v) &= u^T Q v \\ &= \sum_{i,j=1}^{n} u_i q_{ij} v_j \quad . \end{aligned}$$

When $B = D^2 f(x)$ this matrix Q (now depending on x) is the matrix of second partial derivatives of f at x:

$$\begin{pmatrix} \frac{\partial^2 f}{\partial x_1 \partial x_1}(x) & \frac{\partial^2 f}{\partial x_1 \partial x_2}(x) & \ldots & \frac{\partial^2 f}{\partial x_1 \partial x_n}(x) \\ \vdots & \vdots & & \vdots \\ \frac{\partial^2 f}{\partial x_n \partial x_1}(x) & \frac{\partial^2 f}{\partial x_n \partial x_2}(x) & \ldots & \frac{\partial^2 f}{\partial x_n \partial x_n}(x) \end{pmatrix} .$$

This is called the *Hessian matrix* of f at x. It depends on the choice of co-ordinates in $\mathbb{R}^n$, whereas the original bilinear map

$D^2f(x) : E \times E \to R$ does not.

It is well-known that for a point x to be a local *maximum* or *minimum* for f it is necessary that the partial derivatives $\frac{\partial f}{\partial x_1}(x), \frac{\partial f}{\partial x_2}(x), \ldots, \frac{\partial f}{\partial x_n}(x)$ should all vanish, and then in order to determine whether x is a maximum or is a minimum it is necessary to look closely at the properties of the Hessian matrix at x. We shall return to this in detail in §2.9.

It is a standard fact from elementary calculus that when $f : R^n \to R$ is C^2 we always have

$$\frac{\partial^2 f}{\partial x_i \partial x_j}(x) = \frac{\partial^2 f}{\partial x_j \partial x_i}(x)$$

for $i,j = 1,2,\ldots,n$, which means that the Hessian matrix is *symmetric*. This is the same as saying that the bilinear map $D^2f(x) : R^n \times R^n \to R$ is symmetric, i.e.

$$D^2f(x)(u,v) = D^2f(x)(v,u)$$

for all u,v in R^n. More generally, if f is C^r then each rth partial derivative is independent of the order of differentiation. The corresponding co-ordinate-free version of this is that the r-multilinear map $D^rf(x)$ is *symmetric*, i.e. independent of the relative ordering of the r components of $E \times E \times \ldots \times E$. A proof of this fact can be found in Dieudonné [32], for example.

Note in passing that of the two examples of multilinear maps given above, the multiplication example is symmetric but the determinant example is not.

Remarks

1. If $f(x) = (f_1(x), f_2(x), \ldots, f_m(x))$ where x is in R^n then it is straightforward to prove that f is C^r $(r \geq 1)$ precisely when all the rth mixed partial derivatives of all the f_i exist and are continuous. (See Remark 2, §2.4.)

2. An n-multilinear map $E \times E \times \ldots \times E \to R$ is sometimes called a *covariant tensor of rank* n on E. When $E = R^m$ a covariant tensor of rank n is specified by m^n coefficients.

3. In the discussion of multilinear maps we have avoided mentioning topologies for $M_n(E_1 \times E_2 \times \ldots \times E_n;F)$. Strictly speaking we can manage without these since, as we showed, the M_n spaces are just alternative versions of the complicated spaces of continuous linear maps for which at least in principle we already have topologies derived from norms: see §2.3. Nevertheless, it is useful to see how to make M_n into a normed linear space explicitly. By analogy with the theory for continuous linear (= 1-multilinear) maps, it can be shown easily that an n-multilinear map

$$M : E_1 \times E_2 \times \ldots \times E_n \to F$$

is continuous if and only if there exists a constant K such that $||M(x)|| \leq K$ for all $x = (x_1, x_2, \ldots, x_n)$ with $||x_1|| = ||x_2|| = \ldots = ||x_n|| = 1$. If we then take $||M||$ to be the least such K, i.e.

$$||M|| = \sup\{||M(x)|| \;\Big|\; ||x_i|| = 1 \text{ for } i = 1,2,\ldots,n\}$$

then this does define a norm for $M_n(E_1 \times E_2 \times \ldots \times E_n;F)$. What is more, this is in a sense the most reasonable norm to choose, since the natural bijection between M_n and the corresponding space of linear maps

$L(E_1, L(E_2, \dots$ etc.) is then not only a linear isomorphism but is also norm-preserving.

Properties of higher derivatives

Corresponding to the properties of the derivative mentioned in §2.5 we have the following properties of higher derivatives:

(i) Linear combinations of C^r maps are C^r, and

$$D^r(\alpha f + \beta g) = \alpha D^r f + \beta D^r g \quad .$$

(ii) Linear maps L are C^∞, with $D^r L = 0$ for all $r \geq 2$. This is because $DL(x) = L$ for every x, i.e. $DL(x)$ does not vary with x.

(iii) Bilinear maps B are C^∞, with DB as given in §2.5 and D^2B given by

$$D^2B(x)((h_1,h_2),(k_1,k_2)) = B(h_1,k_2) + B(k_1,h_2) \quad .$$

To see this, observe that $DB(x)$ is linear in x and so $D^2B(x) = D(DB)(x) = DB$ for all x. This means that

$$\begin{aligned} D^2B(x)(h,k) &= (D^2B(x).h).k && \text{by definition} \\ &= DB(h).k && \text{by the above} \end{aligned}$$

as claimed. Since $D^2B(x)$ does not vary with x we have $D^rB = 0$ for all $r \geq 3$.

In a similar way it follows that n-multilinear maps M are C^∞, and

$$D^nM(x)(h,k,\dots,p) = \sum_\sigma M(h_{\sigma(1)},k_{\sigma(2)},\dots,p_{\sigma(n)})$$

where there are n entries $h,k,\dots,p$ and the sum is over all permutations σ of $\{1,2,\dots,n\}$. Since this is independent of x it follows that $D^rM = 0$ for all $r \geq n+1$.

(iv) The cartesian product of two C^r maps is C^r, and all derivatives just operate co-ordinate-wise.

(v) The composition of two C^r maps is C^r. This comes from the Chain Rule together with induction on r. We have

$$D(g\cdot f)(x) = Dg(f(x)) \cdot Df(x)$$

so if f and g are C^r then Df and Dg are C^{r-1} and the induction hypothesis (that the result holds for $r-1$) ensures that $Dg(f(x))$ is a C^{r-1} function of x. Then we use the bilinearily of the map

$$L(E,F) \times L(F,G) \to L(E,G) : (A,B) \mapsto B\cdot A$$

and the Chain Rule and induction hypothesis again to see that $D(g\cdot f)$ is C^{r-1} and so $g\cdot f$ is C^r.

There is a formula for $D^r(g\cdot f)$ in terms of derivatives of f and g, but it is cumbersome and not worth trying to memorize.

2.7 GERMS AND JETS

If f and g are two maps which agree on some neighbourhood of a point x then all their derivatives at that point are the same. Therefore if we are interested in trying to deduce the local behaviour of a map from information about its derivatives at x we need not be concerned with the precise nature of the map away from x but could equally well consider any map coinciding with f on some neighbourhood of x. This leads naturally to the idea of the *germ* of a map

Let T be a topological space and let S be any set. Let $f : U \to S$ and $g : V \to S$ be maps with domains U,V open sets in T, and suppose x

lies in $U \cap V$. Then f and g are said to be *germ-equivalent* at x if there exists some open neighbourhood W of x lying inside $U \cap V$ such that $f|W = g|W$, i.e. f and g coincide on W. This is an equivalence relation on the set of all maps defined on neighbourhoods of x in T and with values in S, and the equivalence classes are called *germs* of maps at x. If S is a topological space also then we can consider germs of continuous maps, and if S,T are normed linear spaces E,F we can consider germs at x of C^r maps. We denote the set of these germs by $E_x^r(E,F)$.

In any of these cases, if S is a linear space we can define pointwise addition and scalar multiplication of maps from a given domain into S, and this goes through to the germ level to give the set of germs at x the structure of a linear space. If S also has a multiplicative structure which interacts reasonably with addition and scalar multiplication then this, too, will go through to a similar structure on the set of germs at x. In particular, if S is a *ring* or an *algebra* (whose definitions we will not go into here) then the set of germs at x becomes a ring or an algebra.

The cases which will mainly interest us are those for which $T = R^n$ and $S = R^m$. Taking x without loss of generality to be the origin in R^n, we see that the set $E_0^r(R^n,R^m)$ forms a linear space since R^m does and linear combinations of C^r maps are C^r (property (i) in §2.6). In the case $m = 1$ the codomain R is a ring and even an algebra, and so $E_0^r(R^n,R)$ forms a ring (algebra): here we use properties (iii) and (v) of §2.6 to show that the pointwise product of two C^r functions is C^r. The ring $E_0^\infty(R^n,R)$ of C^∞ germs is often denoted simply E_n or $E(n)$. When studying local properties of C^∞

functions of n variables we will really be studying properties of elements of E_n.

As we have said, if two C^∞ maps are germ-equivalent at x then all their derivatives at x are the same. This means that we can talk about the derivatives at x of a given *germ*, rather than of a specific map representing that germ. The basic question then is: to what extent does knowledge of all the derivatives of a germ at x provide knowledge about the germ itself? Can the germ be reconstructed, given all its derivatives at x?

The most elementary result in this direction is the Mean Value Theorem (§2.5) which for $f : R \to R$ states that

$$f(x+h) = f(x) + hf'(\xi)$$

for some ξ lying between x and h. Of course, this involves $f'(\xi)$ which is not a derivative at x. Therefore

$$f(x+h) = f(x) + hf'(x) + \text{error term}.$$

Further manipulations with the Mean Value Theorem (as in any standard book on calculus) show that the error term can be expressed as $\frac{h^2}{2} f''(\xi_2)$ for some ξ_2 lying between x and $x+h$. Again, this involves a derivative elsewhere than at x, and so if we insist on working with derivatives at x we must write

$$f(x+h) = f(x) + hf'(x) + \frac{h^2}{2} f''(x) + \text{error term}.$$

Once more the Mean Value Theorem can be used to show that the error term will have the form $\frac{h^3}{3!} f''(\xi_3)$ for some ξ_3 between x and $x+h$. This process of pushing the error term further to right can be continued indefinitely, although there is no reason to believe that the error term will become any smaller as this happens. Indeed, it can easily get larger.

The true state of affairs is described by *Taylor's Theorem*, which says that if we write

$$f(x+h) = f(x) + hf'(x) + \frac{h^2}{2!} f''(x) + \ldots + \frac{h^n}{n!} f^{(n)}(x) + \text{error term}$$

then the error term, or *remainder* term, has the form

$$R_n(h) = \frac{h^{n+1}}{n+1!} f^{(n+1)}(\xi_{n+1})$$

for some ξ_{n+1} between x and $x+h$. This $R_n(h)$ is a measure of how inaccurately the first n derivatives of the germ of f at x describe the germ itself. For a given h the remainder term $R_n(h)$ may or may not approach zero as n increases: this must be investigated (or general results appealed to) in any particular case.

Taylor's Theorem has a natural extension to germs of real-valued functions on any normed linear space E

THEOREM (Taylor's Theorem)

Let $f : U \to \mathbb{R}$ *be a* C^∞ *function, where* U *is an open subset of a normed linear space* E*, and let* x *and* $x+h$ *lie in* U *with* h *sufficiently small for the line-segment* ℓ *joining* x *and* $x+h$ *to lie in* U*. Then if we write*

$$f(x+h) = f(x) + Df(x).h + \frac{1}{2!} D^2f(x)(h,h) + \ldots + \frac{1}{n!} D^nf(x)(h,h,\ldots,h) + R_n(h)$$

the error term $R_n(h)$ *has the form*

$$\frac{1}{(n+1)!} D^{n+1}f(\xi)(h,h,\ldots,h)$$

for some point ξ *lying between* x *and* $x+h$ *on* ℓ *.*

Remember that here $D^r f(x)$ is an r-multilinear map

$$E \times E \times \ldots \times E \to \mathbb{R}$$

and so $D^r f(x)(h,h,\ldots,h)$ is a real number. Recall also that when E is euclidean space $\mathbb{R}^n$ we can for purposes of calculations write $D^r f(x)(h,h,\ldots,h)$ as

$$\sum \frac{\partial^r f}{\partial x_i \partial x_j \ldots \partial x_p}(x)\, h_i h_j \ldots h_p$$

where the sum is over all ordered r-tuples $i,j,\ldots,p$ of the integers from 1 to n.

The proof of the general version of the theorem is surprizingly simple, given the one-dimensional version. We write

$$F(t) = f(x + th)$$

and apply the theorem to F at $t = 0$. First of all

$$\begin{aligned} F'(t) &= DF(t).1 \\ &= Df(x + th).h \end{aligned}$$

by the Chain Rule, and then since the map $L(E,\mathbb{R}) \to \mathbb{R}$ taking A to $A.h$ is linear for each fixed h the Chain Rule applied again gives

$$F''(t) = D^2 f(x + th)(h,h)$$

and in general

$$F^{(r)}(t) = D^r f(x + th)(h,h,\ldots,h) \quad .$$

Now writing

$$F(1) = F(0) + F'(0) + \frac{1}{2!}F''(0) + \ldots + \frac{1}{n!}F^{(n)}(0) + \frac{1}{(n+1)!}F^{(n+1)}(\xi)$$

and substituting for $F^{(r)}(0)$ and $F^{(n+1)}(\xi)$ gives us the theorem as stated for f.

If it is the case that $R_n(h)$ does approach zero as n increases then we are, by definition of convergence for infinite series, entitled to write

$$f(x+h) = f(x) + Df(x).h + \frac{1}{2!} D^2f(x)(h,h) + \ldots$$

$$= \sum_{n=0}^{\infty} \frac{1}{n!} D^nf(x)(h,h,\ldots,h) \quad ,$$

where D^0f means f. The infinite series on the right is called the *Taylor series* for (the germ of) f at x.

This definition can be generalized directly to maps $f : E \to F$, since the series itself (now a series of elements in F) does not depend on the Mean Value Theorem. If f is C^∞ the Taylor series always exists, but

(1) it may diverge, or

(2) it may converge, but to something other than $f(x+h)$.

If neither of these happens, and the Taylor series converges to $f(x+h)$ for all h in some neighbourhood of x then f is said to be *analytic* at x. For $E = \mathbb{R}^n$, $F = \mathbb{R}^m$ this means that $f_j(x+h)$ can be expressed for sufficiently small h as a convergent power-series in the co-ordinates h_i, for each component f_j of f. Analyticity is sometimes called *class* C^ω.

Analytic maps are convenient to work with, since we can invoke the whole machinery of differentiation of power-series term by term, inversion of power series, etc., of which details can be found in almost any analysis text. Furthermore, powerful results from the theory of complex functions can also be used. What is perhaps even more significant for practical purposes is the fact that if we replace $f(x+h)$ by

$$f(x) + Df(x).h + \frac{1}{2!} D^2f(x)(h,h) + \ldots + \frac{1}{n!} D^nf(x)(h,h,\ldots,h)$$

which is a *polynomial* in h, then we can be confident that by choosing n

large enough the error in taking this approximation will be insignificantly small.

Despite these enticing advantages of analytic functions, implicitly relied upon in much mathematical modelling in physics, engineering and so on, there are two serious objections - one practical and one philosophical - to working with analytic functions. The practical objection is that functions do arise in real life which look harmless but are in fact *not* analytic even though they are C^∞. A standard example is the function $f : \mathbb{R} \to \mathbb{R}$ defined by

$$f(x) = \left.\begin{array}{ll} e^{-(1/x^2)} & , \quad x > 0 \\ = 0 & , \quad x \leq 0 \end{array}\right\} .$$

This is not analytic at 0, even though it is easy to verify (by induction) that all derivatives at 0 exist and are equal to zero, so the Taylor series at 0 is therefore the zero series, which obviously does not converge to the function. Functions of this kind occur in statistical mechanics and the theory of partial differential equations. The graph of f is as in Figure 12.

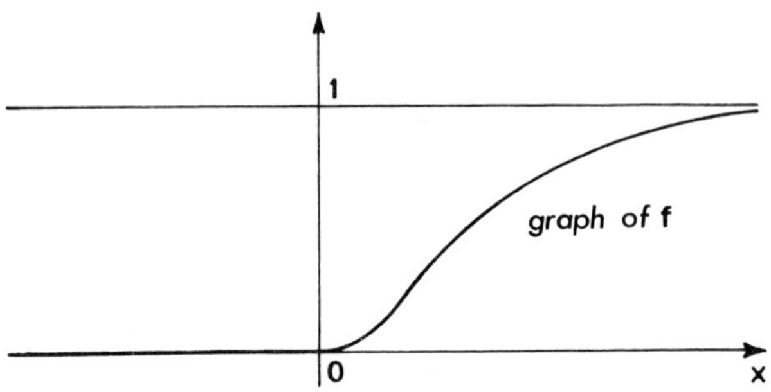

Figure 12

The philosophical objection is that there are no grounds for assuming that smooth functions and maps which model natural forms and processes should share the many very special properties of analytic functions. For example, if the behaviour of an analytic function $f : \mathbb{R} \to \mathbb{R}$ is known on an interval $(-\varepsilon,\varepsilon)$, however small, then it is determined on the whole of $\mathbb{R}$. Related to this is the property that an analytic function cannot be perturbed a small amount locally without also perturbing it globally by what for distant points may be a very large amount. This kind of rigidity is undesirable for models of the real world, and particularly for dynamical systems (Chapter 4) where the idea of stability under small local perturbations plays an important role. Therefore analyticity is not only an unwarranted assumption, but is a constraint that it may be quite inappropriate to impose on a mathematical model.

The technique of representing a function by a finite number of terms in its Taylor series, and then forgetting the rest, is of course basic to many applications of mathematics. To justify this in the case of quantitative calculations it is necessary to make some estimate of the size of the remainder term $R_n(h)$, either in the form in which it appears in the statement of Taylor's theorem above or in one of several other forms in which it can be re-expressed. However, there are many contexts in which precise numerical calculations can be (at least in the first instance, if not ultimately) less important than an understanding of the qualitative way in which the germ of the function behaves. Problems of buckling in structural engineering (see e.g. Thompson and Hunt [141]) and morphogenesis in biology (Thom [138]) are examples of such contexts. In these cases we are not interested so much in the *size* of the remainder term, as in whether by a suitable (possibly non-linear) change in co-ordinates near x the remainder term can be removed completely, so that the function is in a very precise sense equivalent to its truncated Taylor series. This is formalized in the notion of *determinacy*.

First, we introduce some more efficient terminology. Instead of the 'Taylor series of f at x, truncated after the term of degree k' we simply refer to the k-*jet* of f at x, denoted by $j_x^k f$. This can be thought of either as a polynomial in h:

$$f(x) + Df(x).h + \frac{1}{2!} D^2 f(x)(h,h) + \ldots + \frac{1}{k!} D^k f(x)(h,h,\ldots,h)$$

or, more formally, as a collection of multilinear maps

$$\left(f(x), Df(x), \frac{1}{2!} D^2 f(x), \ldots, \frac{1}{k!} D^k f(x)\right) .$$

(Here we are not suggesting that f is multilinear: the first term $f(x)$ in the k-jet at x is the 'constant' map $E \to F$ which takes every h in E to $f(x)$ in F. This is, of course, not the same as f which takes h to $f(h)$.) When $E = \mathbb{R}^n$ and $F = \mathbb{R}^m$ we can choose co-ordinates and represent $j_x^k f$ by an m-vector of polynomials in h whose coefficients are the appropriate mixed partial derivatives of the components of f at x, or we can represent $j_x^k f$ by the coefficients themselves. When $m = n = 1$ this reduces to

$$j_x^k f = (f(x), f'(x), \frac{1}{2!} f''(x), \ldots, \frac{1}{k!} f^{(k)}(x)) .$$

Now we must define what we mean by local equivalence of germs. Let E, F be normed linear spaces, with x a point in E. Recall the term *local diffeomorphism* at x (§2.5), interpreted at least in finite dimensions as a local invertible (non-linear) change in co-ordinates around x.

DEFINITION

Two germs $f, g \in E_x^r(E,F)$ *with* $f(x) = g(x)$ *are*

right equivalent *if there exists a germ* γ *of a local* C^r *diffeomorphism at* x *(in* E*) such that* $g = f\cdot\gamma$ *;*

left equivalent *if there exists a germ* β *of a local* C^r *diffeomorphism at* $f(x)$ *(in* F*) such that* $g = \beta\cdot f$.

Thus by a suitable C^r change of local co-ordinates in E or in F, respectively, the germ f can be transformed into the germ g. Of course we could also define *left-right equivalence*, allowing co-ordinate changes in both domain and codomain simultaneously. It is easy to see that these 'equivalences' are indeed equivalence relations on $E_x^r(E,F)$. See Figure 13.

Finally, we can formulate the meaning of *determinacy*:

DEFINITION

The k-jet ξ *at* x *is* determined *if any two germs at* x *having* ξ *as their* k-jet *are right equivalent.*

In particular, any f with $j_x^k f = \xi$ can be converted by a local change of co-ordinates into ξ thought of as a polynomial. If f is a germ whose k-jet is determined, we say that f is k-*determined*. Sometimes the term k-*sufficient* is also used.

If we have a C^∞ function germ f, and in order to understand the local behaviour of f we write

$$f(x+h) = (f(x) + Df(x).h + \dots + \frac{1}{k!} D^k f(x)(h,h,\dots,h)) + \dots$$

and ignore the final string of dots, then to be sure that nothing

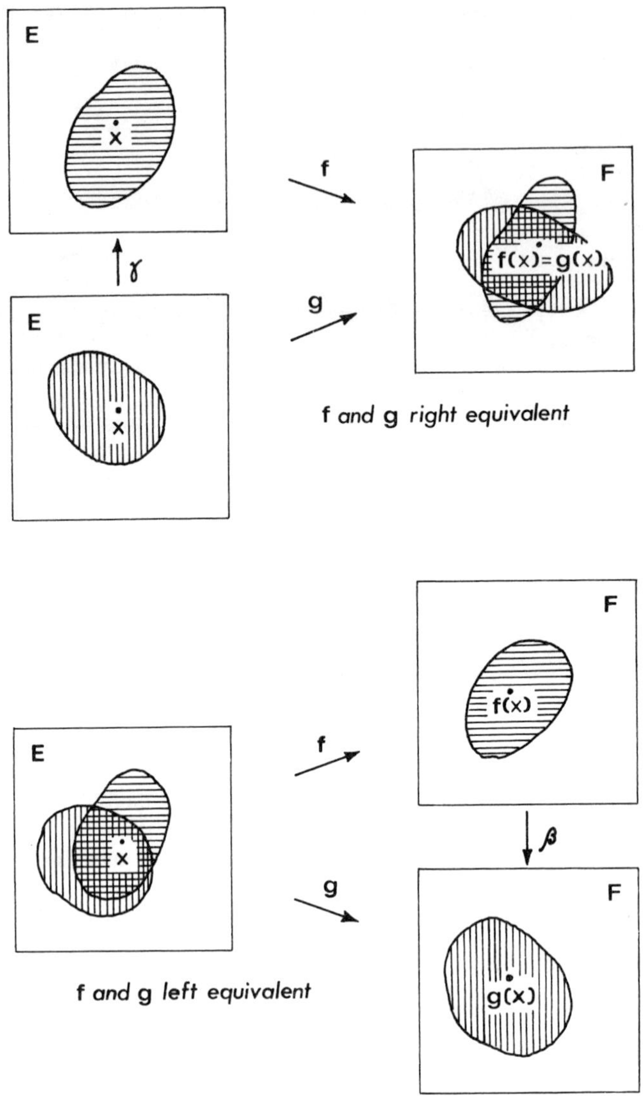
E
x
f
g
F
f(x)=g(x)
γ
E
x
f and g right equivalent
F
f(x)
E
x
f
g
β
F
g(x)
f and g left equivalent

Figure 13

essential has been thrown away we have to know that f is k-determined. For example, the non-analytic function mentioned above cannot be handled near 0 in this cavalier fashion, since every k-jet there is zero and so we obtain no information about even the *sign* of the function for small x (a property independent of choice of co-ordinate) merely by looking at its k-jet. Obviously, this function germ is not k-determined for any k. As an analytic example, consider the germ of f given by

$$f(x,y) = x^2$$

at the origin in R^2. This is not k-determined for any k, since the function

$$g(x,y) = x^2 + y^{2k}$$

(which has the same k-jet as f) is zero at the origin and positive elsewhere, whereas $f(x,y)$ is zero along all the y-axis. Of course, this simple example could be disguised a little by changing co-ordinates and considering $f(x,y) = x^2 + 2xy + y^2$ instead of the original f, and a similar argument would apply relative to the line $x+y = 0$. In view of these discouraging examples we should therefore ask the preliminary question:

(1) are there any k-jets which are determined?

and then, supposing that there are,

(2) what tests are there for deciding whether or not a given k-jet is determined?

We will deal with both these questions in §§2.8, 2.9.

2.8 LOCAL STRUCTURE OF DIFFERENTIABLE MAPS. I: NON-SINGULAR BEHAVIOUR

We have already seen (as a corollary to the Chain Rule) that if $f : U \to V$ is a diffeomorphism and x lies in U then $Df(x) : E \to F$ is an isomorphism, i.e. the invertibility of f as a differentiable map goes over to the invertibility of $Df(x)$ as a linear map. The interpretation of this when $E = F = \mathbb{R}$ is very simple: since $Df(x)$ is an isomorphism precisely when $f'(x) \neq 0$ (remember $Df(x).h = f'(x)h$) the result says that if $f : U \to V$ is a diffeomorphism then $f'(x) \neq 0$ for all x in U. Thus in order for the graph of f over U to be also (when turned on its side) the graph of a differentiable function $V \to U$ it is necessary that its slope should be non-zero everywhere in U. See Figure 14

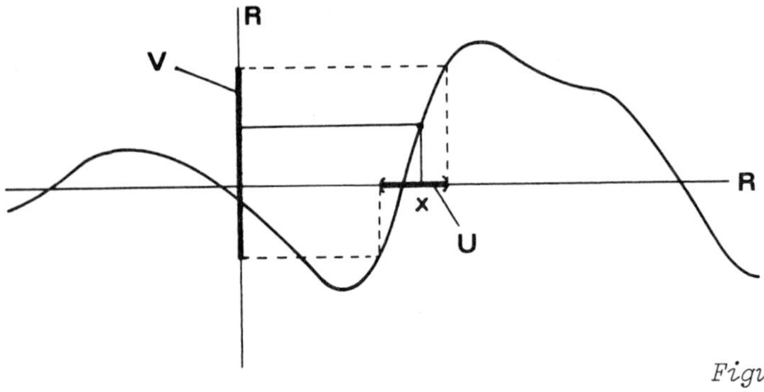

Figure 14

Observe that if $f'(x) = 0$ for some x in U then f fails to be a diffeomorphism on any neighbourhood of x because

either f is not bijective on any neighbourhood of x

or f is bijective on some neighbourhood of x, but f^{-1} is not differentiable at the point $y = f(x)$.

Now it is a standard lemma from elementary analysis that the converse also holds, i.e. $f'(x) \neq 0$ implies that x has a neighbourhood W on which f is a local diffeomorphism. This is often called the Inverse

Function Theorem. It is natural to ask whether its analogue holds in general for maps between normed linear spaces. The answer is that it does, with the one technical qualification that the spaces involved should be complete (see §2.3), i.e. they should be Banach spaces. For finite-dimensional spaces the completeness is no restriction.

THEOREM (Inverse Function Theorem)

Let E, F *be Banach spaces and let* $f : U \to F$ *be a* C^r *map* $(1 \leq r \leq \infty$ *or* $\omega)$*, where* U *is an open set in* E*. Suppose* x *is a point in* U *for which* $Df(x) : E \to F$ *is an isomorphism. Then there is an open neighbourhood* W *of* x *contained in* U *such that the restriction of* f *to* W *is a diffeomorphism from* W *to* $f(W)$.

A proof can be found for example in Dieudonné [32] or Lang [67]. Some texts give the proof only for C^1 maps, but the step from C^1 to C^r is relatively easy.

Writing $Df(x) = A$ in $L(E,F)$, we see that the germs of

$$h \mapsto f(x+h)$$

and

$$h \mapsto f(x) + Ah$$

are *right equivalent*, since we can write

$$\begin{aligned} f(x+h) &= f(x) + AA^{-1}(f(x+h) - f(x)) \\ &= f(x) + A\gamma(h) \end{aligned}$$

where $\gamma(h) = A^{-1}(f(x+h) - f(x))$. Here γ is a diffeomorphism germ since A^{-1} is (because A is, by hypothesis) and the theorem ensures that $h \mapsto f(x+h) - f(x)$ is. In other words, we can *change co-ordinates locally in* $E = R^n$ to *convert* f *into its own linear approximation (1-jet) at* x.

Remarks

1. A continuous linear map $L : E \to F$ between normed linear spaces may be an isomorphism between E and F as linear spaces but the inverse map $L^{-1} : F \to E$ might conceivably fail to be continuous. It can be shown, however, that when E and F are Banach spaces the continuity of L does guarantee the continuity of L^{-1}. Therefore to apply the theorem it is sufficient to check that $Df(x)$ is *bijective*.

2. The proof of the existence of f^{-1} involves an iteration process: this is where the completeness condition comes in. The Mean Value Theorem is also used to give estimates of the discrepancy between f and $Df(x)$.

3. If $E = R^n$ and $F = R^n$ then $Df(x)$ is bijective precisely when *the Jacobian matrix of* f *at* x *is non-singular*. Therefore there is a finite process for checking whether or not f is a local diffeomorphism.

EXAMPLE

Consider the map $\Phi : R^2 \to R^2$ taking (u,v) to $(u \cos v, u \sin v)$, a C^∞ map but obviously not a diffeomorphism since not bijective. The derivative of Φ at (u,v) is represented by the matrix

$$\begin{pmatrix} \cos v & -u \sin v \\ \sin v & u \cos v \end{pmatrix}$$

which is singular precisely where its determinant $u(\cos^2 v + \sin^2 v)$ vanishes, i.e. along the line $u = 0$. Hence, by the Inverse Function Theorem, Φ is a local diffeomorphism in some neighbourhood of each point not on the v-axis. Every $(x,y) \neq (0,0)$ is of the form $\Phi(u,v)$ for some (u,v), so if U denotes the open set $\{(u,v) \mid u \neq 0\}$ in R^2 the restricted map

$$\Phi : U \to \{R^2 \text{ minus the origin}\}$$

is surjective, and locally a diffeomorphism. Writing u,v as r,θ we see that we are simply stating familiar facts about *polar co-ordinates* (see Figure 15):

1. every (r,θ) defines a unique (x,y), but not conversely since $(r,\theta+2n\pi)$ defines the same point as (r,θ);
2. when we restrict θ to avoid ambiguities, the map $(r,\theta) \mapsto (x,y)$ is C^∞ in both directions.

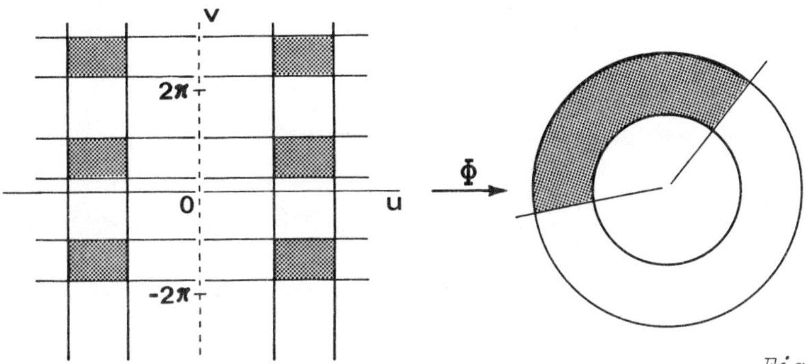

Figure 15

The Inverse Function Theorem is in a sense the key theorem of calculus. In dealing with derivatives of maps (derivative = linear approximation) we lose accuracy but gain linearity, which allows relative ease of calculation. For such calculations to be useful we must know when information about derivatives can be reinterpreted as information about the original map. The Inverse Function Theorem is the first basic result in this direction.

There are two other important results which can be deduced directly from the theorem, although at first sight they look more general. We will state finite-dimensional versions, and comment afterwards on the adjustments needed for infinite-dimensional analogues. Thus we temporarily suppose $E = R^n$, $F = R^m$, and $f : U \to F$ is a C^r map, where U is an open set in R^n. Let x be a point of U, and again write $Df(x) = A$.

COROLLARY 1.

Suppose A *is surjective (so* $n \geq m$*). Then the germs at* 0 *of*

$$h \mapsto f(x+h)$$

and

$$h \mapsto f(x) + Ah$$

are right equivalent.

COROLLARY 2.

Suppose A *is injective (so* $n \leq m$*). Then the germs at* 0 *of*

$$h \mapsto f(x+h)$$

and

$$h \mapsto f(x) + Ah$$

are left equivalent.

Thus Corollary 1 says that if $Df(x)$ is surjective we can change co-ordinates locally in the domain $E = R^n$ to convert f into its own 1-jet, whereas in Corollary 2 we achieve the same effect when $Df(x)$ is injective by changing co-ordinates locally in the codomain $F = R^m$.

Proof of Corollary 1

We can suppose $x = 0$ in R^n and $f(x) = 0$ in R^m, and then we have to show that the germs of f and A are right equivalent at 0. First we number the co-ordinates so that A can be written as a matrix $(B\ C)$ where B is an $m \times m$ non-singular matrix and C is an $m \times (n-m)$ matrix. Simple considerations of bases show that this can be done.

Next we split up R^n correspondingly into factors $R^n = R^m \times R^{n-m}$ with co-ordinates (u,v) and define a new map $F : R^n \to R^n$ by

$$F(u,v) = (f(u,v),v) \quad .$$

Then $DF(0) = L$ has matrix

$$\begin{pmatrix} B & C \\ 0 & I \end{pmatrix}$$

which is non-singular $(I = (m-n) \times (m-n)$ identity matrix), and so by the Inverse Function Theorem F is locally a diffeomorphism. Let γ be the germ at 0 of $F^{-1} \cdot L$. Then

$$\begin{aligned} L(u,v) &= F\gamma(u,v) \\ &= (f\gamma(u,v),\ \text{something}) \end{aligned}$$

giving $f \cdot \gamma = A$ since the first co-ordinate of $L(u,v)$ is $Bu + Cv = A(u,v)$. Hence f is right equivalent to A, proving Corollary 1.

Proof of Corollary 2

Again we suppose $x = 0$ in R^n, $f(x) = 0$ in R^m and we choose co-ordinates such that A has matrix $\begin{pmatrix} B \\ C \end{pmatrix}$ where now B is an $n \times n$ non-singular matrix and C is an $(m-n) \times n$ matrix. We write R^m as $R^n \times R^{m-n}$ with corresponding co-ordinates (x,y) and this time define a map $G : R^m \to R^m$ by

$$G(x,y) = f(x) + (0,y) \quad .$$

Then $DG(0) = M$ has matrix

$$\begin{pmatrix} B & 0 \\ C & I \end{pmatrix}$$

which is non-singular and so by the Inverse Function Theorem G is locally a diffeomorphism. Let β be the germ at 0 of $M \cdot G^{-1}$.

Then

$$M(x,0) = \beta G(x,0)$$
$$= \beta f(x)$$

giving $\beta \cdot f = A$ since $M(x,0) = \begin{pmatrix} Bx \\ Cx \end{pmatrix} = Ax$. Hence f is left equivalent to A, proving Corollary 2. See Figure 16.

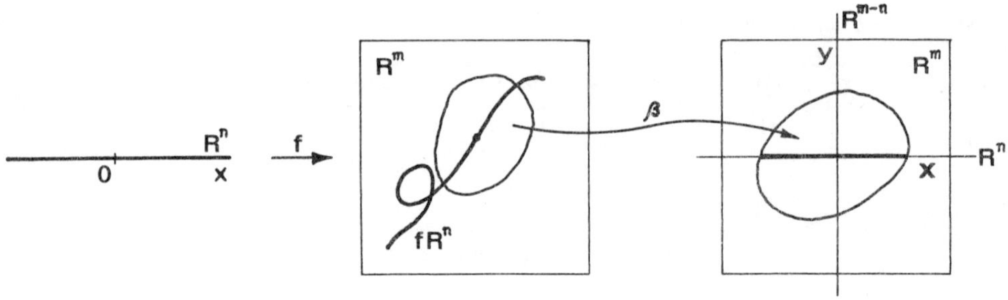

Figure 16

The condition that A is injective means that the kernel of A is zero, i.e. the only n-vector u satisfying $Au = 0$ is the zero vector. This means that the only linear combination $\sum_{i=1}^{n} u_i \underline{c}_i$ of the columns $\underline{c}_i$ of A (as a matrix) which is zero is the one with all $u_i = 0$, or in other words the columns of A are *linearly independent*. Correspondingly, A is surjective precisely when the rows of A are linearly independent. The *rank* of a matrix is the largest number of linearly independent rows, or alternatively, columns (the answer is the same). Thus the two corollaries above can be summarized as follows:

> *If* $Df(x)$ *has maximum rank then* f *looks near* x *like the linear map* $Df(x)$ *near* 0.

Remarks

1. In infinite dimensions we cannot talk about *ranks*, so we must stick to the first formulation of the corollaries. We also have to add in an extra condition to ensure that the splitting of E or F into two suitable 'co-ordinates' (x,y) is still valid: then the proofs are the same as in the finite-dimensional case. Explicitly, the condition is that in Corollary 1 we should be able to split E into a direct sum $E_1 \oplus E_2$ (see page 42) where E_1, E_2 are closed linear subspaces of E with $E_2 = \ker A$ and E_1 taken by A *isomorphically* onto F, or similarly in Corollary 2 we should be able to split F into $F_1 \oplus F_2$ where A takes E *isomorphically* onto F_1. These splitting conditions can be shown to be automatically satisfied when F is finite-dimensional in Corollary 1 or E is finite-dimensional in Corollary 2.

2. In the context of Corollary 1 we can define a C^r map (germ) $\phi : \mathsf{R}^{n-m} \to \mathsf{R}^m$ by

$$\phi(y) = \text{first co-ordinate of } \gamma(-B^{-1}Cy, y) ,$$

where the inspiration for the expression on the right is that the linear subspace of R^n consisting of all elements of the form $(-B^{-1}Cy,y)$ is precisely the kernel of A. It then follows that

$$f(\phi(y),y) = f\gamma(-B^{-1}Cy,y)$$

because γ preserves the y co-ordinate (because L and F do),

$$= A(-B^{-1}Cy,y)$$

$$= 0$$

by design. This version of Corollary 1 is called the *Implicit Function Theorem*, since it says that if

$$f : \mathsf{R}^n = \mathsf{R}^m \times \mathsf{R}^{n-m} \to \mathsf{R}^m$$

is a C^r map with $f(0) = 0$ and such that the derivative of f

restricted to the first factor is an isomorphism (represented by the matrix B above), then there exists a C^r map $\phi : \mathbb{R}^{n-m} \to \mathbb{R}^m$ satisfying

$$f(\phi(y), y) = 0$$

for all y in some neighbourhood of the origin, or in other words putting $f = 0$ gives the first co-ordinate as an implicit function ϕ of the second co-ordinate. Note that when $n = 2$, $m = 1$ we have $f : \mathbb{R}^2 \to \mathbb{R}$ and the derivative condition is simply that $\frac{\partial f}{\partial x} \neq 0$ at the origin.

The Implicit Function Theorem holds for Banach spaces in general, with the proviso about splitting as in Remark 1. In this context it is an extremely important tool, allowing quick proofs of many existence theorems in various branches of non-linear analysis.

In this section we studied the behaviour of smooth maps under conditions of non-singularity of the derivative. The next step is to look at behaviour at points where this fails. Such points are naturally called *singular points*.

2.9 LOCAL STRUCTURE OF DIFFERENTIABLE MAPS. II: SINGULARITIES

We will continue to work entirely in finite dimensions, and comment on infinite-dimensional parallels where appropriate. Let $f : U \to \mathbb{R}^m$ be a differentiable map, where U is an open subset of $\mathbb{R}^n$.

DEFINITION

A singular point *of* f *is a point* x *at which* Df(x) *does* not *have maximal rank, i.e. where* Df(x) *is neither injective nor surjective. The germ of* f *at a singular point is called a* singularity.

Thus at a singular point the derivative may not give a good local qualitative picture of the map itself.

Remark

The term 'singularity' is used in many different fields to mean a variety of different things - such as a point where f is not differentiable, or not continuous, or not defined. It may also refer to a region of space with bizarre geometric properties, such as a 'black hole' in cosmology. Here we will keep the word to mean the germ of a differentiable map at a singular point.

EXAMPLES of singularities

1. In the case $n = m = 1$ the derivative $Df(x)$ is a linear map $R \to R$ which must either be surjective or be the zero map. Thus a singular point of $f : R \to R$ is a point x where $f'(x) = 0$, called a *critical point* or *stationary point*. For example, $f : R \to R$ given by $f(x) = x^2$ has a singularity at $x = 0$ and nowhere else.

2. The previous example generalizes immediately to the case $m = 1$ for any n, i.e. to functions $f : U \to R$. Here the derivative $Df(x)$ is the linear map $R^n \to R$ represented by the $1 \times n$ matrix

$$\left(\frac{\partial f}{\partial x_1}, \frac{\partial f}{\partial x_2}, \ldots, \frac{\partial f}{\partial x_n}\right)$$

and so the only way in which it can fail to have maximal rank $(=1)$ is by the vanishing of all the partial derivatives $\dfrac{\partial f}{\partial x_i}$. Again, this is called a *critical point* or *stationary point* of f. A standard example is $f : R^n \to R$ given by $f(x) = x_1{}^2 + x_2{}^2 + \ldots + x_n{}^2$; this has a singularity at 0 in R^n but nowhere else. As we shall see later, this is in a certain sense the most 'nonsingular' type of singularity.

3. A *constant* map $f : R^n \to R^m$ (i.e. $f(x) = c$ in R^m for all x in R^n) is the most singular kind of map imaginable. Here the derivative $Df(x)$ is the zero linear map $R^n \to R^m$ for each x, and the whole of the structure of R^n is crushed by f into a single point of R^m.

4. As a geometrically more interesting kind of singularity consider $f : R^2 \to R^2$ given by

$$f(x_1,x_2) = (x_1,x_2^3 + x_1x_2) .$$

Here $Df(x)$ is represented by the 2×2 matrix

$$\begin{pmatrix} 1 & 0 \\ x_2 & 3x_2^2 + x_1 \end{pmatrix}$$

and so the rank is less than maximal precisely when

$$3x_2^2 + x_1 = 0$$

which is the equation of a parabola in the domain R^2. The whole map can be described pictorially as in Figure 17.

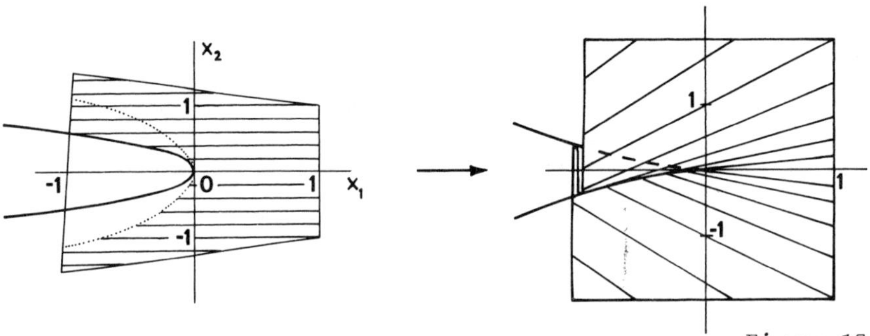

Figure 17

Critical points of real-valued functions

We will now return to the case when $m = 1$, in other words when f is one real function of n variables and singular points mean critical points, i.e. points at which all the partial derivatives of f vanish.

To understand how f behaves near a critical point, it is natural to look at the next available 'approximate' information about f, namely its second derivative. Recall that this second derivative $D^2f(x)$ can be regarded as a symmetic bilinear map

$$D^2f(x) : R^n \times R^n \to R$$

or as a quadratic form, represented by the Hessian matrix

$$H_f(x) = \left(\frac{\partial^2 f}{\partial x_i \partial x_j}(x)\right)$$

of f at x. From experience with the Inverse Function Theorem it seems reasonable that we will be able to extract the most local information about f from its second derivative when this Hessian matrix is non-singular.

DEFINITION

A critical point x *of* $f : U \to R$ *is called* degenerate *or* non-degenerate *according to whether the determinant of the Hessian matrix of* f *at* x *does or does not vanish.*

Remarks

1. The Hessian matrix itself depends on a choice of co-ordinates in R^n, but the vanishing or otherwise of its determinant does not. This is easy to verify by applying a co-ordinate change and then using the Chain Rule twice. However, the real reason for this is that the vanishing or non-vanishing of the determinant of $H_f(x)$ corresponds to the degeneracy or non-degeneracy of the bilinear map $D^2f(x)$.
An arbitrary bilinear map $B : R^n \times R^n \to R$ is said to be *degenerate* if there exists some u in R^n with the property that $B(u,v) = 0$ for every v in R^n ; otherwise it is *non-degenerate*. Thus the critical point x of f is degenerate or non-degenerate according to whether

$D^2f(x)$ is a degenerate or non-degenerate bilinear map.

2. In view of the previous remark, we can extend the theory of critical points to functions on any normed linear space E. The point x is a critical point of f when $Df(x) : E \to R$ is the zero linear map, and is degenerate or non-degenerate according to the degeneracy or non-degeneracy of the bilinear map $D^2f(x) : E \times E \to R$.

The Inverse Function Theorem allowed us to change co-ordinates to bring f locally into linear form, provided the derivative was as well-behaved as possible. We might next hope for a theorem saying that even at a critical point we can change co-ordinates to bring f locally into some agreeable standard form, provided that the critical point is non-degenerate. There is such a theorem. It is called the Morse Lemma, after Marston Morse (b. 1892) who gave it as part of a general theory of the structure of real-valued functions, a theory which has had far-reaching implications in differential topology. Proofs may be found in many books, such as Golubitsky and Guillemin [42] or Milnor [82].

THEOREM (Morse Lemma)

Suppose p *is a non-degenerate critical point of* f. *Then the germ of* f *at* p *is right equivalent to the germ at* 0 *of a function* ϕ *of the form*

$$(x_1,x_2,\ldots,x_n) \mapsto f(p) + \left(\varepsilon_1 x_1^2 + \varepsilon_2 x_2^2 + \ldots + \varepsilon_n x_n^2\right)$$

where each ε_i *is* ± 1 .

It is clear that ϕ does have a non-degenerate critical point at the origin, since $D\phi(x) = (2\varepsilon_1 x_1, 2\varepsilon_2 x_2, \ldots, 2\varepsilon_n x_n)$ (as a $1 \times n$ matrix) which vanishes at $x = 0$, and $H_\phi(0)$ is the $n \times n$ diagonal matrix

$$\begin{pmatrix} 2\varepsilon_1 & & & & \\ & 2\varepsilon_2 & & & \\ & & \cdot & & \\ & & & \cdot & \\ & & & & \cdot \\ & & & & & 2\varepsilon_n \end{pmatrix}$$

which is non-singular. The Morse Lemma says that this is the archetypal example, every other function near a non-degenerate critical point being reducible to one of this type by a local change of co-ordinates.

The number of ε_i which are -1 is called the *index* of the critical point p. It depends only on f near p, and not on the particular way of choosing new co-ordinates; it is the number of negative eigenvalues of $H_f(p)$ or of $D^2f(p)$ as a linear map $R^n \to L(R^n, R) = R^n$.

If the index of p is k we can rearrange co-ordinates so that ϕ has (apart from the constant $f(p)$) the form

$$(x_1, x_2, \ldots, x_k) \mapsto \left(x_1^2 + \ldots + x_{n-k}^2\right) - \left(x_{n-k+1}^2 + \ldots + x_n^2\right) .$$

This allows us to gain a good geometric picture of the behaviour of ϕ near 0, i.e. of f near p. There are $n + 1$ possible values of the index (including $k = 0$) and $n + 1$ correspondly different types of non-degenerate critical point.

$\underline{n = 1}$	$k = 0$	Local form x^2	: minimum
	$k = 1$	Local form $-x^2$	: maximum
$\underline{n = 2}$	$k = 0$	Local form $x_1^2 + x_2^2$	: minimum
	$k = 1$	Local form $x_1^2 - x_2^2$	: saddle point
	$k = 2$	Local form $-x_1^2 - x_2^2$	: maximum

See Figure 18 for pictures in the case $n = 2$. Note that a saddle point can be thought of as a minimum in one direction coupled with a maximum in another. In general there will be $(n-1)$ different types of saddle point corresponding to $k = 1,2,\ldots,n-1$.

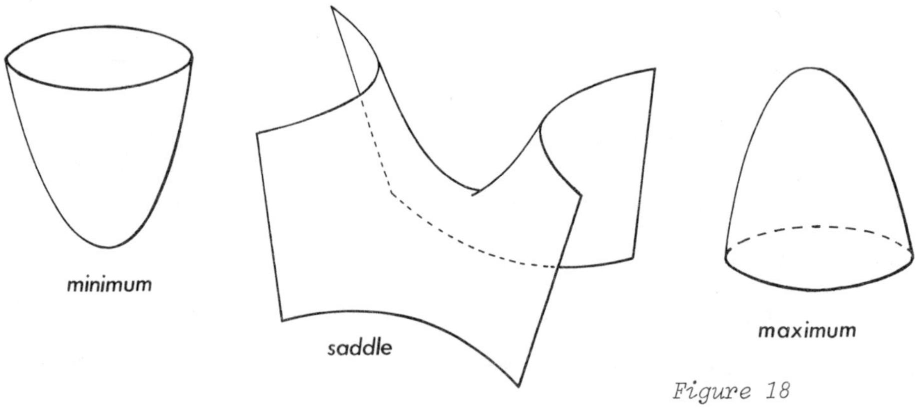

Figure 18

Remarks

1. An immediate deduction from these local forms is that each non-degenerate critical point is *isolated*, i.e. lies in some neighbourhood containing no other critical point, degenerate or otherwise.

2. The Morse Lemma has an infinite-dimensional version, extremely useful in calculus of variations where it is necessary to consider the behaviour of a function defined on a normed linear space E whose elements are themselves functions. See Palais [92]. In finite dimensions we could write the (rearranged) canonical form for ϕ as

$$x \mapsto f(p) + ||(I-P)x||^2 - ||Px||^2$$

where P is projection onto the last k co-ordinates and I denotes the identity map $R^n \to R^n$. The infinite-dimensional Morse Lemma states that, provided E is a Hilbert space (page 53), in a neighbourhood of a non-degenerate critical point p the function f can be converted into the above form where P now denotes a perpendicular projection onto some linear subspace of E. There do exist generalizations of this to Banach spaces which are not Hilbert spaces: see Tromba [144].

Further study of degenerate critical points

At a degenerate critical point p of f the Hessian matrix $H_f(p)$ is singular, which is equivalent to saying that as a linear map $R^n \to R^n$ the matrix has non-zero kernel, i.e. there exists at least one direction u with $H_f(p).u = 0$. However, it would be reasonable to hope that if we keep perpendicular to such directions we could still apply a kind of Morse Lemma giving a standard quadratic form in some co-ordinates, leaving a completely degenerate (i.e. degenerate in all directions) function of the remaining variables. Such a generalized Morse Lemma, allowing us to choose co-ordinates so as to split the function locally into the sum of a non-degenerate 'Morse' part and a completely degenerate part, conveniently does exist. It was first published in an infinite-dimensional context by Gromoll and Meyer [45].

THEOREM (Splitting Lemma, or Gromoll-Meyer Lemma)

Suppose p *is a degenerate critical point of* f. *Then the germ of* f *at* p *is right equivalent to the germ at* 0 *of a function* ϕ *of the form*

$$(x_1, x_2, \ldots, x_n) \mapsto f(p) + \left(\varepsilon_1 x_1^2 + \ldots + \varepsilon_r x_r^2\right) + \psi\left(x_{r+1}, \ldots, x_n\right)$$

where each ε_i *is* ± 1 *and* ψ *is completely degenerate, i.e.* $D^2\psi(0) = 0$ *(as well as* $D\psi(0) = 0$*).*

Remark

In the infinite-dimensional version when E is a Hilbert space H we let K be the kernel of the second derivative $D^2f(p)$ regarded as a linear map $H \to L(H,R) = H^*$. (In fact for a Hilbert space $H^* = H$, in much the same way that $R^{n*} = R^n$; see page 247.) We assume that K is finite-dimensional, so that there is only a 'finite-dimensional amount' of degeneracy, and that the image of $D^2f(p)$ is a closed subspace of H^* (a technical necessity). Each point x can be regarded as having co-ordinates (w,y) where y belongs to K and w is perpendicular to K. Then the standard form given by the Lemma is

$$x \mapsto f(p) + ||(I-P)w||^2 - ||Pw||^2 + \psi(y)$$

where this time P is a projection inside the space of vectors w perpendicular to K, and ψ satisfies $D\psi(0) = 0$ and $D^2\psi(0) = 0$ as an element of $L(K,K^*)$ or as a bilinear map $K \times K \to R$.

In view of the Gromoll-Meyer Lemma, the problem of studying the behaviour of degenerate critical points has been narrowed down to the study of critical points at which the second derivative vanishes entirely, i.e. points at which the 2-jet vanishes (ignoring the initial constant $f(p)$). It would be pleasant if there were some kind of 'non-degeneracy' condition on the third derivative in the presence of which we could again convert the function locally into some standard form by a sort of third degree Morse Lemma. Essentially this is the case, but the details of the theory, which generalizes to functions with vanishing 3-jets, 4-jets, etc., are rather more subtle than one might at first suspect. Indeed, it is only in recent years with the work of Thom, Mather, Arnol'd and others that the classification of degenerate critical points under various conditions of relative non-degeneracy has become properly understood at all.

The problem can be stated in two parts:

(1) To what extent does the first non-vanishing k-jet of f determine the local behaviour of f ?

(2) Can the first non-vanishing k-jet (which is a homogeneous polynomial in $x_1, x_2, \ldots, x_n$ of degree k) be put into some normal form?

The second problem is one of algebraic geometry, and can be solved more or less easily for a few low values of k and n. For example, when $k = 2$ we have the classification of quadratic forms into standard non-degenerate types

$$\varepsilon_1 x_1^2 + \varepsilon_2 x_2^2 + \ldots + \varepsilon_n x_n^2 \quad (\varepsilon_i = \pm 1)$$

which appeared in the Morse Lemma, and degenerate types

$$\varepsilon_1 x_1^2 + \varepsilon_2 x_2^2 + \ldots + \varepsilon_r x_r^2 \quad (\varepsilon_i = \pm 1,\ 0 \leqslant r < n)$$

which featured in the Gromoll-Meyer Lemma. When $k = 3$ and $n = 2$ a cubic form in x_1, x_2 factorizes into one or three real factors (which may coincide) giving without much effort the various standard forms

$$x_1^3 + x_2^3,\ x_1^3 - 3x_1 x_2^2,\ x_1^2 x_2,\ x_1^3 .$$

The first problem is that of *determinacy* of the k-jet (see §2.7). Without going into the depths of the theory, we will give the result in the form of a finite procedure for testing the determinacy of the first non-vanishing jet.

We may as well suppose that we are working at the origin in R^n, and we shall regard the first non-vanishing jet of a function germ as being a homogeneous polynomial of the appropriate degree k. Let ξ be such a k-jet, and for $i = 1,2,\ldots,n$ let d_i denote $\frac{\partial \xi}{\partial x_i}$, itself either zero or a homogeneous polynomial of degree $k - 1$.

DETERMINACY TEST. *The jet* ξ *as above is determined if and only if every homogeneous polynomial* q *of degree* $k+1$ *in the* x_i *can be written as*

$$q = q_1 d_1 + q_2 d_2 + \dots + q_n d_n$$

where the q_i *are homogeneous polynomials of degree* 2.

Since the space of homogeneous polynomials of degree k is a finite-dimensional real linear space of dimension $\frac{(n+k-1)!}{(n-1)!k!}$ (the elements $x_1^{m_1} x_2^{m_2} \dots x_n^{m_n}$ with $m_1 + m_2 + \dots + m_n = k$ form a basis), the Determinacy Test reduces to a finite combinatorial problem.

EXAMPLES

1. The k-jet of a constant function is not determined since each $d_i = 0$. However, this fact is obvious anyway.
2. If $Df(0) \neq 0$ (i.e. 0 is not a critical point of f) then the 1-jet ξ of f is determined since at least one of the d_i (say d_1) is a non-zero number and then every polynomial q of degree $k+1 = 2$ can be written as d_1 times a polynomial of degree 2. However, we already knew this fact as a consequence of the Inverse Function Theorem in §2.8.
3. If $Df(0) = 0$ then if A denotes the Hessian matrix $H_f(0)$ we have

$$d_i = \sum_{j=1}^{n} A_{ij} x_j \quad .$$

If A is non-singular (i.e. 0 is a non-degenerate critical point of f) there is a matrix $B = A^{-1}$ such that

$$x_j = \sum_{i=1}^{n} B_{ji} d_i \ .$$

Now every homogeneous polynomial q of degree 3 can be written as $q = \sum_{j=1}^{n} x_j p_j$ where the p_j are homogeneous of degree 2, and so

$$q = q_1 d_1 + q_2 d_2 + \dots + q_n d_n$$

where $q_i = \sum_{j=1}^{n} B_{ji} p_j$ is homogeneous of degree 2. Thus we deduce that the 2-jet of f is determined. We already knew this fact by the Morse Lemma.

4. The k-jet $x \mapsto x^k$ is determined, since every homogeneous polynomial in x of degree $k+1$ is simply of the form cx^{k+1} which can be written as $(ck^{-1}x^2)\frac{d\xi}{dx}$. Although we have not met this result before, it is not particularly surprizing. It merely asserts that any function which can loosely be thought of as

$$x^k + (\text{higher order terms})$$

can be converted into y^k near 0 by a suitable local change of co-ordinates $y = y(x)$. To do this informally we would write the function of x as

$$x^k(1 + \phi(x))$$

where $\phi(0) = 0$, and then take $y = x(1 + \phi(x))^{1/k}$. It is easy to make this argument rigorous.

Observe in constrast:-

5. For $k > 1$ the k-jet $(x_1,x_2) \mapsto x_1^k$ is *not* determined, since x_2^{k+1} cannot be written as

$$q_1 \frac{\partial\xi}{\partial x_1} + q_2 \frac{\partial\xi}{\partial x_2} = q_1 k x_1^{k-1} \quad .$$

The result is reasonable, since x_1^k vanishes everywhere along a line (the x_2-axis) whereas no amount of local changes of co-ordinates around 0 in R^2 will alter the fact that e.g. $x_1^k - x_1^{k-1}x_2^2$ vanishes also along a curve touching the x_2-axis at the origin.

More generally, it is easy to see both from the algebra and the geometry that no k-jet of $f : R^n \to R$ which is independent of one or more of the co-ordinates can possibly be determined.

6. The 3-jet

$$(x_1,x_2) \mapsto x_1^3 - 3x_1x_2^2$$

is determined since $d_1 = 3x_1^2 - 3x_2^2$, $d_2 = -6x_1x_2$ and every homogeneous polynomial of degree 4 in x_1,x_2 can be written as $q_1d_1 + q_2d_2$ with q_1,q_2 of degree 2. To see this, observe that $x_1^3x_2$, $x_1^2x_2^2$, $x_1x_2^3$ are already multiples of d_2 and that

$$x_1^4 = \frac{1}{3}x_1^2d_1 - \frac{1}{6}x_1x_2d_2$$

$$x_2^4 = -\frac{1}{3}x_2^2d_1 - \frac{1}{6}x_1x_2d_2 \quad .$$

However:

7. The 3-jet

$$(x_1,x_2) \mapsto x_1^2x_2$$

is *not* determined, since $d_1 = 2x_1x_2$, $d_2 = x_1^2$ and so every combination of d_1 and d_2 would have x_1 as a factor. Thus x_2^4 is not of the form $q_1d_1 + q_2d_2$. This non-determinacy can be understood geometrically as follows: $x_1^2x_2$ vanishes along two lines (the co-ordinate axes) cutting at right angles, whereas $x_1^2x_2 + x_2^5$, for example, vanishes along the x_1-axis only. No local co-ordinate changes in R^2 can convert one of these configurations into the other.

So far we have considered determinacy only for *homogeneous* k-jets, which materialize as the first non-vanishing jet at 0 of some function germ. It is important also to be able to test determinacy of non-homogeneous jets, since it frequently happens that although the first non-vanishing jet of f may not be determined, the addition of the next term in the Taylor series of f gives a jet which then is determined. The necessary and sufficient test for determinacy of non-homogeneous jets is a little cumbersome, and in many cases it is enough to work with two tests, one necessary and one sufficient, which do not quite meet in the middle. For full details see Poston and Stewart [101].

Let ξ be any (possibly non-homogeneous) k-jet, and define $d_i = \frac{\partial \xi}{\partial x_i}$ as before. In this case d_i may be a non-homogeneous polynomial of any degree $\leq k - 1$. Let the *minimum degree* of a polynomial denote the lowest degree of all its terms. For example, the minimum degree of $x_1^2 x_2 x_3 + x_1^3 x_3^2$ is four.

Sufficient test. *If every homogeneous polynomial* q *of degree* k + 1 *can be written as*

$$q = \sum_{i=1}^{n} q_i d_i + r \qquad (*)$$

where the q_i *are polynomials of minimum degree at least* 2, *and* r *is a polynomial of minimum degree at least* k + 2, *then* ξ *is determined.*

This is similar to the earlier test (for homogeneous jets), but we now allow the q_i to be non-homogeneous and discard all terms of degree $> k + 1$ which arise.

EXAMPLE

We saw previously that $(x_1,x_2) \mapsto x_1^2x_2$ is not a determined 3-jet.

If now we let η denote the 4-jet $(x_1,x_2) \mapsto x_1^2x_2 + x_2^4$ we find that

$$x_1^5 = -2x_1^2x_2^2d_1 + x_1^3d_2$$

$$x_2^5 = -\frac{1}{8}x_1x_2d_1 + \frac{1}{4}x_2^2d_2$$

and other homogeneous polynomials of degree 5 have d_1 as a factor and so η is determined (here $r \equiv 0$). Thus the undetermined 3-jet $x_1^2x_2$ has been made into a determined 4-jet by adding on x_2^4.

Necessary test. *If* ξ *is determined then every homogeneous polynomial of degree* $k+1$ *can be written as* (*) *where the* q_i *are polynomials of minimum degree* 1 *and* r *is as before.*

Clearly the sufficient test implies the necessary test, as it should do. However, there is a small gap between the two through which jets can easily fall.

EXAMPLE (Siersma's example [117])

Let ξ be the 4-jet $(x_1,x_2) \mapsto x_1^3 + x_1x_2^3$. We have $d_1 = 3x_1^2 + x_2^3$, $d_2 = 3x_1x_2^2$. A little combinatorial work shows there is no way in which x_2^5 can be written as $q_1d_1 + q_2d_2$ (modulo terms of degree $\geqslant 6$) with q_1, q_2 each of minimum degree $\geqslant 2$. Thus the Sufficient Test fails and so we suspect that ξ may not be determined. To prove this we must apply the Necessary Test: here we find that any homogeneous polynomial of degree 5 *can* be written as $q_1d_1 + q_2d_2$ with q_1, q_2 each of minimum degree 1 (in particular we see $x_2^5 = x_2^2d_1 - x_1d_2$) and so ξ does not fail the Necessary Test and we cannot be sure that it is not determined: it has fallen between the two tests. In fact this ξ *is* determined, but to prove this needs more technicalities which we will not discuss here.

Interestingly, for this ξ it turns out that every homogeneous polynomial of degree 6 can be written as $q_1d_1 + q_2d_2$ (modulo terms of degree 7) where q_1, q_2 have minimum degree 2. This means that ξ *regarded as a* 5-*jet* is determined, in other words that any function germ of the form

$$(x_1, x_2) \mapsto x_1^3 + x_1x_2^3 + (\text{terms of degree} \geqslant 6)$$

is right equivalent to ξ. It is in showing that terms of degree 5 also make no difference that the subtlety of this example lies.

Remarks on the literature. There are a great many books on linear algebra. A fairly random selection is Cullen [31], Halmos [49], Mirsky [84] (in which linear spaces are called linear *manifolds*), Nomizu [91], Shields [114]. The basic material on calculus in normed linear spaces can be found in Dieudonné [32], Lang [67] or Loomis and Sternberg [70], and for euclidean space in almost any text on advanced calculus such as Hoffman [58], Rudin [109] or Spivak [128]. See also Griffiths and Hilton [43] or Hirsch and Smale [55] for both linear algebra and calculus. Some references on singularity theory are Eells [36], Golubitsky and Guillemin [42], or the survey by Arnol'd [9]. Determinacy of jets is explored and explained thoroughly in Poston and Stewart [101]. Much of the content of this chapter is also contained in books on differential topology referred to at the end of Chapter 3, and also in Field [160].

3 Differentiable manifolds and maps

3.1 THE CONCEPT OF A DIFFERENTIABLE MANIFOLD

With the topology and calculus developed so far we have assembled a mathematical toolkit for dealing with

1. differentiation in open subsets of euclidean spaces or, more generally, in normed linear spaces
2. global structures which are not necessarily linear.

The next step is to unite these two. As a natural result of the synthesis of *local linearity* with *global non-linearity* we arrive at the definition of a differentiable manifold.

Recall that in order to formulate the definition of *differentiability* at x for a continuous map $f : U \to F$, where U is open in E, we only needed to use the linearity of E, F *locally*, i.e. in some neighbourhood of x in U and $f(x)$ in F. Therefore if S and T are two topological spaces we can talk about differentiability of a map $f : S \to T$ at a point x in S provided that S and T look locally like normed linear spaces in neighbourhoods of x and $f(x)$. Now 'looks like' in the context of topological spaces means 'is homeomorphic to', so we can express all this more formally as follows.

Suppose S, T are topological spaces and $f : S \to T$ is a continuous map. Let x be a point of S, and suppose that x has an open neighbourhood U in S such that there is a homeomorphism $\phi : U \to U'$ where U' is an open subset of a normed linear space E. Suppose also that $f(x)$ has an open neighbourhood V in T with a homeomorphism

$\psi : V \to V'$ where V' is an open subset of a normed linear space F, and assume for simplicity that U was chosen small enough in the first place so that $f(U) \subset V$ (this is possible since f is continuous). Then the composition $\psi \cdot f \cdot \phi^{-1}$, which we shall write simply as $\psi f \phi^{-1}$, is a continuous map from the open subset U' of E to the normed linear space F. (Its codomain is strictly V', but then we just include V' in F.) See Figure 19. This map $\psi f \phi^{-1}$ 'represents' f locally as the kind of map with which we can do calculus as in §2.4.

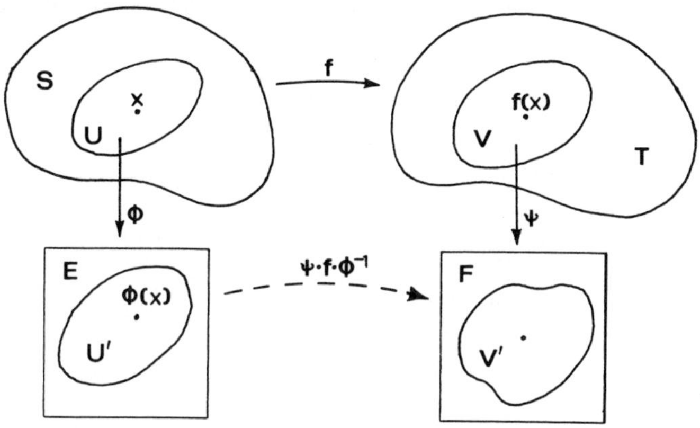

Figure 19

In general we will be interested in studying the properties of f not just at one point x of S but over the whole of S, and so we will want S to be covered by a family of open sets U_α, each one equipped with a homeomorphism

$$\phi_\alpha : U_\alpha \to U'_\alpha$$

from U_α onto some open set U'_α in a normed linear space E_α. The same applies to the space T. If we put in the further reasonable assumption that S looks locally the same everywhere, i.e. that all the E_α's are really just copies of the same normed linear space E, we have arrived at the definition of a *manifold* modelled on E.

DEFINITION

A manifold *modelled on a normed linear space* E *is a topological space* S *which admits a family* $\{U_\alpha\}$ *of open sets covering* S *such that each* U_α *is homeomorphic with an open subset* U'_α *of* E.

Each U_α, together with a particular homeomorphism $\phi_\alpha : U_\alpha \to U'_\alpha$, is called a *chart*. The collection $\{(U_\alpha, \phi_\alpha)\}$ of all the charts is an *atlas* for S.

In the case when E is *finite-dimensional* we can choose coordinates in each U'_α, and thus establish (via ϕ_α) coordinates on each subset U_α of S. For this reason charts are often called (local) *coordinate charts*. If $f : S \to T$ is a map between two manifolds then we can express local representatives $\psi f \phi^{-1}$ in terms of local coordinates and do calculus locally as if we were in euclidean space, representing derivatives by matrices and so on. We will return to this later.

EXAMPLES of manifolds

1. S = E itself, or any open subset of E.

Here we take an atlas consisting of exactly one chart: the set S itself with the identity map as homeomorphism $\phi : S \to S \subset E$.

2. S = any open subset of a manifold T.

Take an atlas $\{(U_\alpha, \phi_\alpha)\}$ for T, and observe that its restriction $\{(S \cap U_\alpha, \phi_\alpha | S \cap U_\alpha)\}$ to S forms an atlas for S.

3. S = the circle $S^1 = \{(x_1, x_2) \in R^2 | x_1^2 + x_2^2 = 1\}$.

There are plenty of convenient atlases. For example:

(a) Four open sets U_1, U_2, U_3, U_4 defined as those points (x_1, x_2) in S^1 for which

$$x_1 > 0, \quad x_1 < 0, \quad x_2 > 0, \quad x_2 < 0$$

respectively, together with corresponding homeomorphisms ϕ_i from U_i each to the open interval $(-1, 1)$ in R given by

$$\phi_1(x_1,x_2) = \phi_2(x_1,x_2) = x_2$$

$$\phi_3(x_1,x_2) = \phi_4(x_1,x_2) = x_1 \ .$$

(b) Two open sets U_+, U_- defined as those points (x_1,x_2) in S^1 for which

$$x_2 \neq -1, \quad x_2 \neq +1$$

respectively, together with homeomorphisms

$$\phi_\pm : U_\pm \to R$$

given by

$$\phi_\pm(x_1,x_2) = x_1/(1\pm x_2) \ .$$

The simplest way to see that $\phi_\pm$ are homeomorphisms is to draw a picture. The formulae for $\phi_\pm$ express the fact that $\phi_\pm$ is stereographic projection from the sphere minus the point $(0, \mp 1)$ onto the x_1-axis thought of as a separate copy of R: see Figure 20.

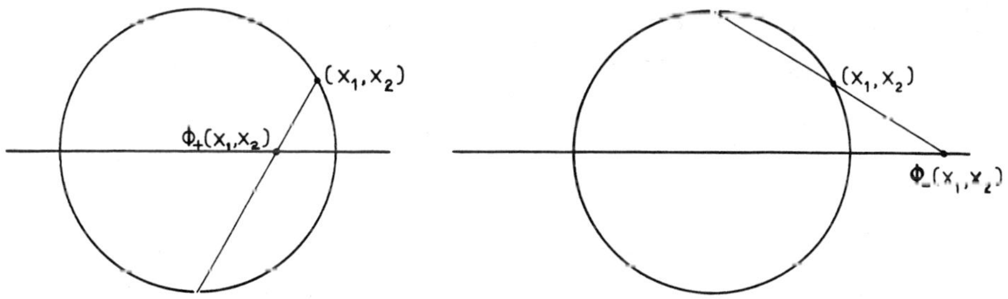

Figure 20

Drawing a diagram is valuable in (a) too. For both (a) and (b) it is easy to write down the details to prove that the maps used are indeed homeomorphisms.

4. S = the n-sphere $S^n = \left\{(x_1,x_2,\ldots,x_{n+1}) \in R^{n+1} \;\middle|\; \sum_{i=1}^{n+1} x_i^2 = 1\right\}$.

The ideas behind the atlases (a), (b) used for S^1 extend easily to apply to the n-sphere S^n for any n. To understand this it is worthwhile working out corresponding formulae explicitly for the (easily visualizable) two-sphere S^2.

5. The subset S of R^2 consisting of the union of the two coordinate axes is *not* a manifold, since every neighbourhood of the origin 0 in S contains a set like $+$ which is not locally like any R^n at 0. The proof of this plausible claim follows by considering what is left when 0 is removed: the complement of 0 in $+$ falls into four disjoint non-empty open subsets, while the complement of any connected neighbourhood of a point in R^n either falls into two open subsets (n=1) or remains connected (n > 1) when the point is removed (cf. §1.7). These descriptions are entirely topological and so invariant under homeomorphism, showing that 0 has no neighbourhoods in $+$ homeomorphic to open subsets of any R^n.

6. S = the square perimeter $\{(x_1,x_2) \mid \max(|x_1|, |x_2|) = 1\}$ in R^2. At first sight there seems to be something wrong here since a square has corners, which seems incompatible with our general programme of taking local coordinates everywhere in which to do (smooth) calculus. However, the definition of manifold is satisfied by this S, since S is homeomorphic to the circle S^1 (via radial projection from the origin, for example) and so an atlas for S^1 will provide one for S also. This illustrates the general fact that *anything homeomorphic to a topological manifold is itself a topological manifold,* since if $h : T \to S$ is a homeomorphism and $\{(U_\alpha,\phi_\alpha)\}$ is an atlas for S then $\{(h^{-1}U_\alpha, \phi_\alpha \cdot h)\}$ is one for T.

Despite the formal manoeuvring in this last example, we may still feel uncomfortable about trying to do calculus around the edge of a square. We would be right: our instincts detect the fact that manifolds as we have defined them do not yet quite provide the correct setting for global calculus. We will pursue this problem further, after ending the list above with two rather bizarre examples of manifolds.

7. $S = R^2$ with an unusual topology constructed as follows:

> U is open in S precisely when U intersects every line ℓ parallel to the x_2-axis in an open subset of ℓ (regarded as R with the usual topology).

Thus every 'usual' open subset of R^2 is open in this topology, but certainly not conversely. For example, the x_2-axis is open in S but not in the usual R^2. Every point p in S has a neighbourhood homeomorphic to R, namely the line through p parallel to the x_2-axis. Thus S is a manifold modelled on R, not on R^2.

This manifold is in a certain sense embarrassingly large, as it possesses an *uncountably infinite* number of disjoint open sets. It needs an uncountably infinite number of charts to constitute an atlas.

8. $S = [0,\infty)$ with an unusual topology as follows:

> the open sets in S are precisely those which are open in the usual topology (induced from R) and which are either disjoint from 0 or contain 0 together with an open interval $(1, 1+\varepsilon)$ for some $\varepsilon > 0$.

It is easy to verify that this prescription does define a topology for S. Furthermore, every point has neighbourhoods homeomorphic to open intervals in R : this is by definition for points other than 0, and a little checking reveals that each neighbourhood of 0 in S of the form $[0,\varepsilon) \cup (1,1+\varepsilon)$ is homeomorphic to $(-\varepsilon,\varepsilon)$ in R. Thus S is a manifold modelled on R. However, S is not Hausdorff, since every neighbourhood of 0 meets every neighbourhood of 1.

There are contexts in which the study of 'large' manifolds and non-Hausdorff manifolds is important, but we shall not be seriously concerned with them. From now on all our manifolds will be assumed automatically to be *Hausdorff* and *possess atlases with at most a countably infinite number of charts*.

Next we look a little more closely at what is involved in doing calculus on manifolds. As already described, we discuss differentiability of maps between manifolds by representing the maps locally (via charts) as maps between open subsets of normed linear spaces. It is clear, though, that the result of this discussion may depend heavily on the choices of the two chart homeomorphisms ϕ and ψ used in the representation. Specifically, if x lies in both U_α and U_β then (for a given ψ) we might find that the representative $\psi f \phi_\alpha^{-1}$ for f was differentiable at $\phi_\alpha(x)$ but the representative $\psi f \phi_\beta^{-1}$ was not differentiable at the corresponding point $\phi_\beta(x)$.

To see how to set about resolving this difficulty, first observe that since x lies in both U_α and U_β we can write the map

$$\psi f \phi_\beta^{-1} : \phi_\beta(U_\alpha \cap U_\beta) \to V' = \psi V \subset F$$

as

$$\phi_\alpha \phi_\beta^{-1} : \phi_\beta(U_\alpha \cap U_\beta) \to \phi_\alpha(U_\alpha \cap U_\beta)$$

followed by

$$\psi f \phi_\alpha^{-1} : \phi_\alpha(U_\alpha \cap U_\beta) \to V'$$

or in other words as a composition:

$$\psi f \phi_\beta^{-1} = (\psi f \phi_\alpha^{-1}) \cdot (\phi_\alpha \phi_\beta^{-1}) \quad .$$

See Figure 21. One representative for f is therefore the

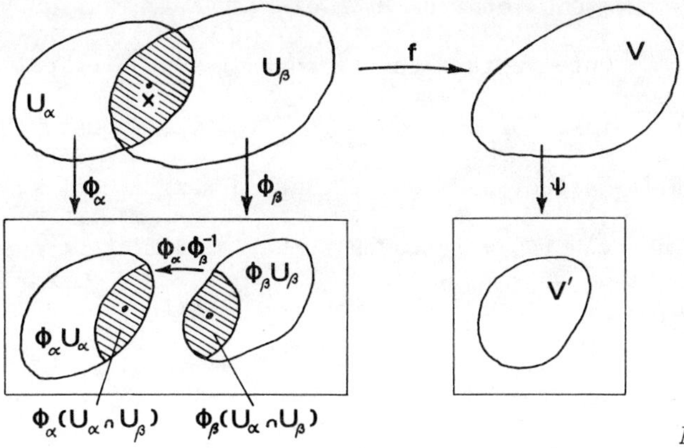

Figure 21

other representative composed with an *overlap map* $\phi_\alpha\phi_\beta^{-1}$. Note that any such overlap map is a homeomorphism of one open subset of E (the model space) with another: we have $(\phi_\alpha\phi_\beta^{-1})^{-1} = \phi_\beta\phi_\alpha^{-1}$.

If it happened to be the case that *every overlap map within the given atlas were of class* C^r $(1 \leq r \leq \infty, \omega)$ then by the Chain Rule (§§2.5, 2.6) we could immediately deduce from the above that $\psi f\phi_\alpha^{-1}$ would be of class C^r precisely when $\psi f\phi_\beta^{-1}$ was of class C^r, thus removing in one stroke all the ambiguity about C^r differentiability arising from the possible choices of different local representatives for f. This assumption about differentiability of the overlap maps within the atlas is the extra ingredient we need to do global calculus satisfactorily.

By working with a given collection of charts we may still lack flexibility. For example, we might be given some coordinate charts on a manifold representing the states of evolution of a particular dynamical system, and wish to use some coordinates different from these in order to express the equations of the dynamics in a more convenient form. In fact we want to be free in principle to use *all* local C^r changes of coordinates that are theoretically possible, everywhere on the manifold. Therefore before we start to work with a manifold we should ideally equip ourselves with *all possible additional charts which overlap differentiably* (C^r) *with the given ones.* Once we take this important psychological step we are thinking no longer in terms of one explicit (finite or countable) atlas but of the 'maximal' atlas associated to it. Such a maximal atlas, which will necessarily have an uncountably infinite number of different possible charts but all overlapping (where they meet) C^r differentiably, is called a C^r *structure* for the manifold. A manifold with a given C^r structure is called a C^r *differentiable manifold.*

We summarize all this by repeating the key definitions.

DEFINITION

An atlas $A = \{(U_\alpha, \phi_\alpha)\}$ *for a manifold is called a* C^r atlas $(1 \leqslant r \leqslant \infty, \omega)$ *if for every* α, β *with* $U_\alpha \cap U_\beta$ *non-empty the overlap map*

$$\phi_\alpha \phi_\beta^{-1} : \phi_\beta(U_\alpha \cap U_\beta) \to \phi_\alpha(U_\alpha \cap U_\beta)$$

is of class C^r *(as a map between open subsets of the model space* E *).*

A C^r *atlas which is maximal (i.e. is already deemed to contain all charts which overlap* C^r *differentiably with its own charts) is called a* C^r structure *for the manifold.*

A C^r differentiable manifold *is, formally, a pair* (M, A) *where* M *is a manifold which admits a* C^r *atlas and* A *is a particular choice of* C^r *structure for* M.

Usually we think of M as being the differentiable manifold, with the C^r structure A understood. A manifold by itself, with no C^r structure being considered at all, is often called a *topological manifold* to emphasize the lack of any additional (differentiability) assumptions beyond purely topological ones. Observe that a C^r structure is a C^s atlas for every $s \leq r$, and hence defines a C^s structure (the associated maximal C^s atlas).

Retracing the steps that led up to this definition, we now see that if M,N are two C^r differentiable manifolds it makes sense to talk about differentiability of continuous maps between M and N : differentiablity is a local property, and so we use local representatives and know that it does not matter which particular local representatives we use.

DEFINITION

A continuous map $f : M \to N$ *between two* C^r *differentiable manifolds is called a* C^r map *if all its local representatives via charts of* M *and* N *are* C^r *maps in the sense of* §2.6.

Note that to verify this in any given case it suffices to check the differentiability of the local representatives with respect to just one convenient atlas for each of M and N. It then follows automatically for all other atlases in the same C^r structure.

As a special case of the above we have:

DEFINITION

A homeomorphism $f : M \to N$ *which is a* C^r *map and whose inverse* $f^{-1} : N \to M$ *is also a* C^r *map is called a* C^r diffeomorphism.

If there exists a diffeomorphism between M and N they are said to be *diffeomorphic* (with the qualifier C^r inserted according to context). Two diffeomorphic manifolds are indistinguishable as far as their topologies and differentiable structures are concerned.

EXAMPLES of differentiable manifolds

All the examples of manifolds above are in fact examples of differentiable manifolds, since the overlap maps of the charts are all C^∞. For example, in 3(a) the overlap map $\phi_3\phi_1^{-1}$ has domain the open interval $(0,1)$ and image also $(0,1)$ (observe that $U_1 \cap U_3$ consists of those points of S^1 lying in the first quadrant) and is given by the formula $t \mapsto +\sqrt{1-t^2}$ which is C^∞ as $0 < t < 1$. The other overlap maps are equally easy to check. In fact the atlases 3(a) or 3(b) define the *same* C^∞ structure for S^1, since the charts from one overlap C^∞ with the charts from the other and therefore a maximal C^∞ atlas including either of them would include both.

Example 6 (the square) still seems worrying. How can a differentiable manifold have sharp corners? Yet the square *is* a differentiable manifold since the overlap map between $\phi_\alpha \cdot h$ and $\phi_\beta \cdot h$ is $(\phi_\beta \cdot h)(\phi_\alpha \cdot h)^{-1}$ which is nothing other than $\phi_\beta \cdot \phi_\alpha^{-1}$, which could perfectly well have been arranged to be C^∞ as in the examples for S^1 above.

The point here is that the 'differentiable' in 'differentiable manifold' refers to a *relative* property of overlap maps, and so if you live in the world defined everywhere locally by the charts you simply do not see the sharp corners. The corners exist only as a feature of the square *in* something else, such as the plane R^2. Once the manifold is considered as part of something bigger then it is having to carry more than its own intrinsic properties as a manifold, and it is important to distinguish these extrinsic properties (whatever they may be in any given context) from the intrinsic ones which do not vary with the context.

What, then, is the real nature of the extrinsic property of the circle/

square manifold that has corners, or, conversely, what is the meaning of not having corners? A straight line in R^2 clearly has no corners, and nor does anything which can be converted into a straight line by a diffeomorphism. Generalizing this to k-dimensional *affine* subspaces of R^n (i.e. linear subspaces translated by some constant vector, so not necessarily passing through the origin in R^n) and realizing that having corners or otherwise is a *local* property (although extrinsic) we arrive at the definition of a subset S of R^n having 'no corners'. A subset of this kind is called a *submanifold* of R^n.

DEFINITION

A submanifold *of* R^n *is a subset* S *of* R^n *with the following property: for each* x *in* S *there is a neighbourhood* U *of* x *in* R^n *and a diffeomorphism* $\phi : U \to U'$, *where* U' *is an open set in* R^n, *such that* $\phi(S \cap U) = L \cap U'$ *where* L *is some affine subspace of* R^n.
See Figure 22.

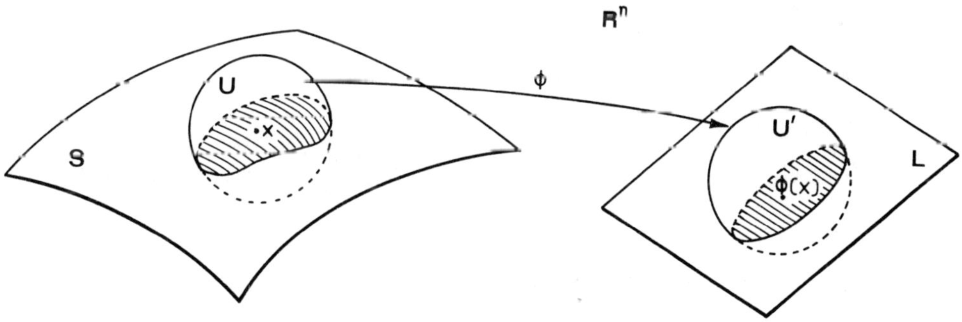

Figure 22

Remarks

1. Officially we should call S a *differentiable submanifold*, or C^r submanifold for some r.

2. Provided S is connected, the dimension of L has to be the same for all x (since for each k the set of x for which $\dim L = k$ is an open subset of S). Therefore by changing ϕ we could take L to be the same for all x, and may as well take it to be the copy of R^k in R^n corresponding to the first k coordinates. Then we could also assume ϕ chosen so that $\phi(x)$ is the origin. This is the form in which the definition is often given. Note that it means that only the first k coordinates in the chart (U,ϕ) are relevant for describing anything that happens in S. Informally we think of S as a k-dimensional hypersurface in R^n.

3. The idea generalizes easily to infinite dimensions, where we replace R^k in R^n by F in E where F is a linear subspace of E. It is usually desirable to assume that F is *closed* as a subset of E, and that E splits into a direct sum $E = F \oplus G$ for some linear subspace G of E (cf. Remark 1, page 99) in order to prove results about submanifolds analogous to finite-dimensional ones that we shall see later. The splitting is automatic in finite dimensions.

Now, referring back to the (intrinsic) definition of differentiable manifold, we observe that *any submanifold of* R^n *is a differentiable manifold in its own right*. To prove this we have to exhibit an appropriate atlas, but this is very easy to do. For each x in S we take the neighbourhood $U \cap S$ of x in S (S having the topology induced from R^n) where U is as in the definition of submanifold, and as chart map we take the restriction ϕ^0 of ϕ to $U \cap S$. If ϕ has been chosen as in Remark 2 we see that $\phi^0 : U \cap S \to U' \cap R^k$ is a homeomorphism from a neighbourhood of x in S to a neighbourhood of 0 in R^k, and so S is a (topological) manifold. Overlap maps for this atlas on S are simply the overlaps of pairs of diffeomorphisms ϕ defined on various open sets in R^n, but with information about the last n-k coordinates discarded: hence they are as smooth as ϕ. Thus a C^r submanifold of R^n inherits from R^n a natural structure of a C^r differentiable manifold.

EXAMPLES of submanifolds of R^n

1. Let $S = S^1$ in R^2. Given any $x = (x_1, x_2)$ on S there are local diffeomorphisms in R^2 converting S locally near x into a straight line segment in R^2. For example, let U be the half-strip $\{x \in R^2 \mid x_1 > 0, |x_2| < 1\}$ in R^2, and define $\phi : U \to R^2$ by

$$(x_1, x_2) \mapsto (x_1 - \sqrt{1-x_2^2}, x_2) \quad .$$

This is a diffeomorphism onto its image $U' = \{x \in R^2 \mid x_1 > -\sqrt{1-x_2^2}, |x_2| < 1\}$ and takes $U \cap S$ to the line segment $\{x \in R^2 \mid x_1 = 0, |x_2| < 1\}$. See Figure 23. It is easy to find

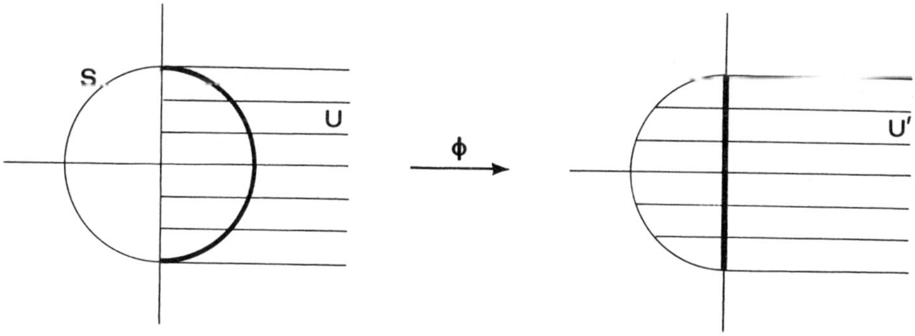

Figure 23

charts (U,ϕ) to take care of points on S with $x_1 < 0$ or $x_1 = 0$ (use x_2 instead), thus showing that S is a C^∞ submanifold of R^2. Note that the restriction of ϕ to $U \cap S$ above is simply the chart map ϕ_1 in Example 3(a), when R is thought of as the x_2-axis in R^2, and indeed any chart for S arising from its submanifold property in R^2 would, if it were not already one of those in 3(a), certainly overlap C^∞ with them. This means that the C^∞ structure for $S = S^1$ defined by the atlas 3(a) (or 3(b)) is the *same* C^∞ structure that S inherits as a submanifold of R^2.

2. Likewise it can be seen that S^n is a C^∞ submanifold of R^{n+1}, although later we shall find a much easier way of showing this than by brandishing atlases explicitly. This C^∞ structure for S^n $(n \geq 1)$ is usually called its *standard* structure.

Now we can unravel the confusion surrounding the 'square' example. The fact is that the square S is not a differentiable *submanifold* of R^2, since at any corner there is no local diffeomorphism in R^2 which will straighten out S locally to a straight line. Nevertheless S can be equipped with a C^∞ structure by exploiting its topological equivalence to a circle as on page 126, but this structure bears little direct relationship to the C^∞ structure of R^2. Examples like this hardly ever arise in practical situations, since manifolds are very often encountered initially as submanifolds of R^n. The only reason for mentioning it was to highlight the distinction between intrinsic and extrinsic properties of manifolds.

3. Suppose we wish to study the motion of a spherical pendulum. The *position* is represented by a point x in the unit 2-sphere S^2, say, and

the *velocity* is represented by a vector v in R^3 which is tangent to the sphere at x, which is the same as saying that v is perpendicular to x regarded as a vector in R^3. Thus the space we must consider if we wish to analyze the global dynamics of the pendulum is the space

$$T = \left\{ (x,v) \in S^2 \times R^3 \;\middle|\; v \text{ is perpendicular to } x \right\}$$

which consists of all elements $(x_1,x_2,x_3;\ v_1,v_2,v_3)$ that satisfy

$$x_1^2 + x_2^2 + x_3^2 = 1$$

and

$$x_1v_1 + x_2v_2 + x_3v_3 = 0 \ .$$

This set T is a submanifold of R^6, as can be proved either directly by a careful choice of charts or indirectly by more general methods: see §3.4 below.

An obvious question arising is: can every differentiable manifold be realized as a submanifold of some R^n? If this is so, then we can in theory dispense with the abstract definition. The answer is Yes: see §3.2, [7] below. However, the ways of doing this may be quite artificial, and distract from the essential properties of the manifold itself. For this reason we prefer to deal with manifolds abstractly where possible, to avoid carrying around a lot of redundant information.

3.2 REMARKS, COMMENTS, AND MORE EXAMPLES OF DIFFERENTIABLE MANIFOLDS

[1] A manifold is n-*dimensional*, or an n-*manifold*, if the normed linear space on which it is modelled is a finite-dimensional space of dimension n. In this case we can regard the model space as R^n. An *infinite-dimensional manifold* is one which is modelled on an infinite-dimensional normed linear

space.

If $\phi : U \to U'$ is a chart on an n-manifold we can for each point u in U write

$$\phi(u) = (\phi_1(u), \phi_2(u), \ldots, \phi_n(u)) \quad \text{in } R^n.$$

Sometimes it is formally useful to write x_i to denote the function $\phi_i : U \to R$ (the ith coordinate function) as well as the ith coordinate itself of x in R^n. If u also lies in the domain V of another chart map ψ then similarly writing y_j to denote ψ_j $(j = 1,2,\ldots,n)$ we can express the fact that the overlap map $\psi\phi^{-1}$ is a C^r diffeomorphism by:

all partial derivatives of the y's *as functions of the* x's *exist (and are continuous) up to order* r, *and the Jacobian matrix*

$$\frac{\partial(y_1, y_2, \ldots, y_n)}{\partial(x_1, x_2, \ldots, x_n)}$$

is non-singular everywhere where it is defined.

This is a consequence of the Inverse Function Theorem (§2.8).

If everything that we have done so far is re-cast in the setting of *complex linear spaces*, with complex differentiability (see Remark 3, page 60) we obtain *complex manifolds*. These arise in the global theory of functions of one or more complex variables, in particular physics for example.

[2] Any open subset W of a C^r manifold M is itself a manifold of the same dimension with charts the restrictions to W of those in the C^r structure of M. In particular (as already noted in §3.1) any open subset of the model space E is a C^∞ manifold. Observe that the meaning of *diffeomorphism* (page 125) is in this case the same as that originally used in §2.5.

3 We have so far avoided asking an obvious question: given a topological manifold is it always possible to choose an atlas whose overlap maps are C^1, thus making the manifold into a differentiable manifold? The answer is No, although not surprizingly the proof involves esoteric topological methods. An example in dimension 10 was found by Kervaire [64] in 1960. However, it can be shown that if there is a C^1 atlas then there will also exist a C^∞ atlas and even a C^ω atlas. Therefore from now on we will tend to work with C^∞ differentiable manifolds, and call them *smooth manifolds*. Remarks made about smooth manifolds will usually apply equally well in the C^r case for any given $r \geqslant 1$.

4 Another natural question to ask is: can a given manifold be equipped with two distinct smooth structures? It is very easy to see that the answer is Yes. For example, consider the two single-chart atlases on $\mathbb{R}$

(1) $\phi : \mathbb{R} \to \mathbb{R} : x \mapsto x$

(2) $\psi : \mathbb{R} \to \mathbb{R} : x \mapsto x^3$.

Officially, these define distinct C^1 structures since their overlap map $\psi\phi^{-1}$ although a differentiable homeomorphism is not a diffeomorphism (its inverse is not differentiable at 0), and therefore $(\mathbb{R},\psi)$ does not belong to the C^1 structure determined by $(\mathbb{R},\phi)$. Nevertheless, the two copies of $\mathbb{R}$ are diffeomorphic as smooth manifolds. The map

$$f : \mathbb{R}_{(1)} \to \mathbb{R}_{(2)}$$

taking x to $\sqrt[3]{x}$ is a diffeomorphism because its representative via the given charts is $\psi f \phi^{-1}$ which is a diffeomorphism of the model space $\mathbb{R}$ – the identity map, in fact.

Therefore a more sensible question to ask is: can a manifold have two distinct smooth structures which are not diffeomorphic? The uncomfortable answer is Yes, although as in 3 it needs sophisticated topological machinery to prove this. For example, in 1956 Milnor [81] showed that there exist smooth structures on a 7-sphere which make it not diffeomorphic to the standard structure on S^7 as a submanifold of R^8. These smooth manifolds are called *exotic spheres*. Fortunately the smooth structure of euclidean space R^n *is* unique up to diffeomorphism, although the proof is of course difficult.

[5] Once we have a few examples of smooth manifolds such as S^n, R^n and open subsets of them, we can build up further examples from them. For instance, if M,N are manifolds of dimensions m,n respectively then their cartesian product $M \times N$ (see §1.3) is an (m+n)-manifold inheriting a smooth structure from smooth structures on M and N. Charts (U,ϕ) on M and (V,ψ) on N define a chart

$$\phi \times \psi : U \times V \to R^m \times R^n = R^{m+n},$$

and it is easy to check that all overlap maps of these charts are C^∞. Here we are saying little more than that if $(x_1,x_2,\ldots,x_m)$ are local coordinates on M and $(y_1,y_2,\ldots,y_n)$ are local coordinates on N then $(x_1,x_2,\ldots,x_m, y_1,y_2,\ldots,y_n)$ are local coordinates on $M \times N$.

Simple examples are the *torus* $S^1 \times S^1$, the infinite *cylinder* $S^1 \times R$, the *solid torus* $S^1 \times D$ where $D = B_1(0)$ is the open unit disc in the plane, and various higher-dimensional analogues such as $S^m \times S^n$, $S^m \times R^n$, $S^1 \times S^1 \times \ldots \times S^1$ and so on. The torus can be visualized in R^3 as a surface of revolution obtained by rotating a circle about a line that lies in the plane of the circle but does not meet the circle. Such an object is a smooth submanifold of R^3, diffeomorphic to the more abstractly defined torus $S^1 \times S^1$: a diffeomorphism is in effect a choice of circular coordinates (θ,ϕ) on the surface. Similarly the cylinder can be seen as a subset of R^3, and is diffeomorphic to the plane R^2 minus the origin via the map

$$S^1 \times R \to R^2 - \{0\}$$

$$(\theta,t) \mapsto (e^t\cos\theta,\ e^t\sin\theta),$$

where θ denotes an angle modulo 2π. Since the interval (a,b) is diffeomorphic to $(0,\infty)$ for any a,b (via $s \mapsto (s-a)/(b-s)$ for example) it follows that $S^1 \times R$ is diffeomorphic to an *annulus* $A = \{x \in R^2 \mid a < ||x|| < b\}$. The solid torus $S^1 \times D$ is diffeomorphic to the open subset of R^3 consisting of points inside a toral surface of revolution, hence the name.

6 A submanifold of R^n was defined in §3.1 as a subset with particularly agreeable local properties, namely that it was everwhere locally like R^k in R^n for some fixed k. However, local phenomena in R^n transfer directly to local phenomena in arbitrary n-manifolds, via the charts, provided they are phenomena that depend only on the smooth structure of R^n (and not, for instance, on the specific linear structure). Therefore the definition of submanifold in R^n lifts immediately to give a definition of k-*dimensional submanifold of* M for any smooth n-manifold M : a subset S of M is a k-dimensional smooth submanifold of M if every point of S lies in some chart (U,ϕ) for M with ϕ taking $S \cap U$ to $R^k \cap U'$ where $U' = \phi U$. See Figure 24. A submanifold of M is itself a smooth

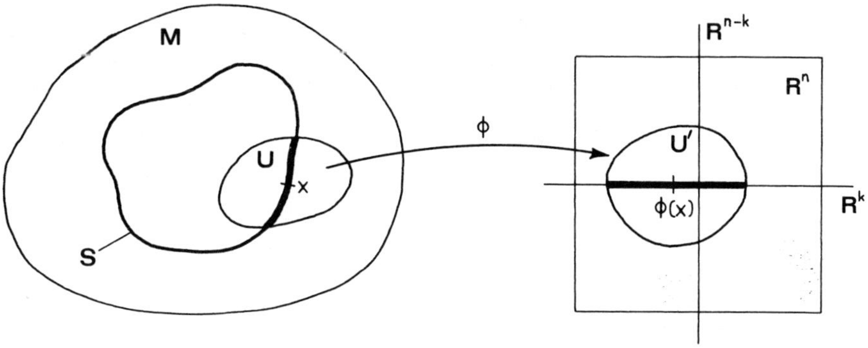

Figure 24

manifold, just as in the case $M = R^n$. Again, it is only the first k coordinates in the chart (U,ϕ) that are relevant to describe what happen in S : the other coordinates are *ignorable*.

An open subset of a manifold M is a submanifold, with $k = n$. For obvious reasons it is called an *open submanifold* of M. If S is a k-dimensional submanifold of M and also an open set in M then k must equal n.

It is often convenient to work not with the dimension k but with the *codimension* $c = n-k$ of a submanifold. Codimension even makes sense when both S and M are infinite-dimensional, if the model space for M is the product of the model space for S and the finite-dimensional space R^c. From the Inverse Function Theorem, this will be the case if every point p on S has an open neighbourhood U in M for which $S \cap U = f^{-1}(0)$ where $f:U \to R^c$ is a submersion (see pp. 153-4). Submanifolds of finite codimension in infinite-dimensional manifolds are important in bifurcation theory: see §4.7.

7 Another natural question we have already touched on in §3.1 is: can every abstractly defined manifold by realized as (i.e. shown to be diffeomorphic to) some submanifold of some euclidean space R^m ? The answer is Yes, and it is not particularly hard to prove this (see e.g. Wallace [145]). What is more difficult to establish is the minimum dimension of euclidean space in which a given n-manifold can be realized. It was shown by Whitney [147] in 1936 that every n-manifold can be realized in R^{2n+1}, and he later [148] improved this to R^{2n}. There exist n-manifolds which cannot be realized in R^{2n-1} (see below for some 2-manifolds which

will not go into R^3), but in many particular cases it is possible to lower the figure $2n$ quite considerably.

In view of this it is not logically necessary to bother with the abstract definition of manifold at all, since every manifold can be regarded as a submanifold of some euclidean space. Nevertheless, as already indicated, it is important to separate the intrinsic from the extrinsic, the essential from the inessential, when dealing with objects as potentially complicated as manifolds. Moreover, a manifold may have physical significance as the global arena, locally coordinatized, within which a dynamical system evolves; a realization in a larger euclidean space may have no physical meaning. The Universe itself can be modelled as a 4-dimensional differentiable manifold: in which R^n are we to embed it?

8 Both in the study of the geometry of manifolds and in the use of manifolds as the natural setting for dynamical systems it can be useful to think of 'cutting' an n-manifold along some (n-1)-dimensional submanifold slice. Locally this is like cutting R^n along R^{n-1}, which we can consider as resulting in two objects

$$H^n_+ = \{x \in R^n \mid x_n \geqslant 0\}$$

$$H^n_- = \{x \in R^n \mid x_n \leqslant 0\}$$

which are to be glued together along

$$R^{n-1} = \{x \in R^n \mid x_n = 0\}$$

to reconstitute R^n. This local picture leads us to define an n-*manifold with boundary* to be a topological space in which every point has open neighbourhoods homeomorphic to open sets in H^n_+ (or H^n_-, but then H^n_+ and H^n_- are homeomorphic) with the topology induced as a subset of R^n. Note that open sets in H^n_+ come in two types: those that contain points of

R^{n-1} and those that do not. Correspondingly, points of an n-manifold with boundary come in two types: (a) those which have genuine R^n-euclidean neighbourhoods, and (b) those whose neighbourhoods look like 'halves' of neighbourhoods in R^n, i.e. neighbourhoods of points of R^{n-1} in H^n_+. The set of all the points of type (b) is called the *boundary* ∂M of the original manifold-with-boundary M. It forms a manifold of dimension $n-1$, and itself has no boundary. We could formally regard any manifold (as defined in §3.1) as a manifold-with-boundary having empty boundary but pedantry would only cause confusion so we usually reserve the 'with boundary' clause for cases when the boundary is really there.

There is no problem in defining C^r structures for manifolds with boundary, provided the obvious steps are taken to define differentiability for maps defined on open sets in H^n_+ rather than just in R^n.

EXAMPLES

1. The 2-disc $D^2 = \{x \in R^2 \mid x_1{}^2 + x_2{}^2 \leqslant 1\}$ is a 2-manifold with boundary. Its boundary ∂D^2 is the circle $S^1 = \{x \in R^2 \mid x_1{}^2 + x_2{}^2 = 1\}$. This example immediately extends to n dimensions, giving $S^{n-1} = \partial D^n$.
2. Examples can be manufactured from products (cf. [5]). The product $S^1 \times [0,1]$ is a cylinder, a compact 2-manifold with boundary whose boundary consists of the two circles $S^1 \times \{0\}$ and $S^1 \times \{1\}$, i.e. the ends of the cylinder. This illustrates that ∂M need not be connected.
3. A manifold with boundary could arise as the setting for a dynamical system in which the variables were subject to some constraint expressed either locally in terms of (weak) inequalities in local coordinates or globally in terms of a bound on energy or some other function. We will mention this again briefly in §4.7, although generally in Chapter 4 we shall

be dealing with systems on manifolds without boundary.

9 Associated to the plane R^2 is the fundamental notion of *orientation*. A coin heads uppermost cannot be changed to tails uppermost simply by sliding it around in the plane. Because of this we can describe a diffeomorphism between connected open sets in R^2 as being either *orientation-preserving* or *orientation-reversing*, according to whether it takes heads to heads or heads to tails - somewhat distorted of course. The reason for assuming the sets are connected is that otherwise the diffeomorphism might preserve orientation on one component and reverse it on the other. This heads/tails notion can be made precise topologically, but it is easier to exploit differentiability and hence (via derivatives) linearity to formalize it. A linear isomorphism $L : R^2 \to R^2$ is *orientation-preserving* or *-reversing* according to whether its determinant is positive or negative; it is easy to verify that this does correspond to the heads/tails idea. By analogy, we define a linear isomorphism $R^n \to R^n$ to be *orientation-preserving* or *-reversing* according to the sign of its determinant. Then we define a diffeomorphism $U \to U'$ to be orientation-preserving if its derivative at each point is orientation-preserving, and finally we define a smooth manifold to be *orientable* if it possesses an atlas in which all overlap maps are orientation-preserving diffeomorphisms. In two dimensions this means, roughly speaking, that we can be 'heads uppermost' or 'tails uppermost' in a consistent way over the whole manifold, and in n dimensions the idea is the same although instead of two sides of a coin we have to think of something like the even-ness or odd-ness of permutations of n local coordinate directions.

A connected manifold which is not orientable is *non-orientable*. All the explicit examples we have considered so far are orientable. Here are some which are non-orientable.

EXAMPLES of non-orientable manifolds

1. *The Möbius strip.* Take a strip of paper, give it a twist of 180°, and glue the ends together: see Figure 25. This is a paper model of a non-orientable 2-manifold with boundary, known as the *Möbius strip* or *band* (A.F. Möbius, 1790-1868). In a sense the Möbius strip is the archetypal

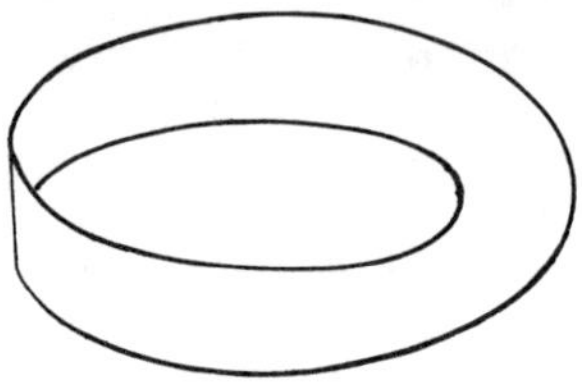

Figure 25

non-orientable 2-manifold, since it can be shown that any non-orientable 2-manifold must contain one. Observe that with two theoretical coins on the strip it is impossible to say whether or not they are the 'same way up', since the result depends on the route you take to slide one to the position of the other to compare them: see Figure 26. On a sphere or a torus it

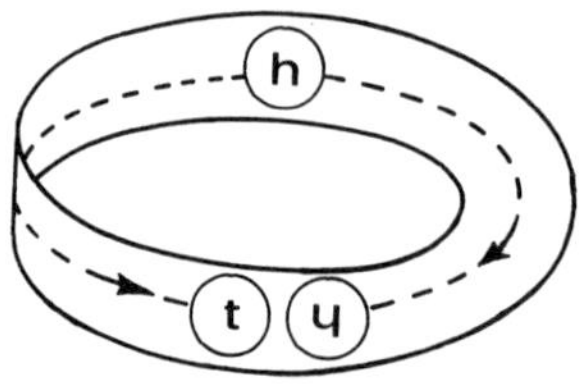

Figure 26

would not matter which route you took: you would always get the same result.

As defined above, the Möbius strip is a piece of paper and not a mathematical object. The mathematical construction goes as follows. Let A be the square $[0,1] \times [0,1]$, and let R be the equivalence relation that relates every point to itself (automatically) and otherwise relates the point $(0,x)$ to the point $(1,1-x)$ for every x in the second factor $[0,1]$. The Möbius strip is then the quotient space A/R (see §1.3). The natural map $A \to A/R$ corresponds to the act of glueing two opposite sides of the square A together in opposite senses, i.e. twisted through 180°. Here A/R inherits a smooth structure from that of the plane in which A lies.

2. *The projective plane* P^2. This is usually defined as the set of all lines through the origin in R^3, topologized as follows: a set L of lines is *open* precisely when the union of all the lines in L, with the origin removed, is an open subset of R^3 with its usual topology. Every such line meets the unit sphere in just two antipodal points, and so P^2 can be described equivalently as the set of all pairs of antipodal points in S^2, with a subset of these pairs being *open* precisely when it defines a pair of antipodal open sets in S^2. More concisely, P^2 is the quotient space S^2/R where R is the equivalence relation 'lying on the same line through the origin'. The standard smooth structure on S^2 gives one on P^2 (take an atlas of charts (U_α, ϕ_α) on S^2 small enough so that no U_α contains two antipodal points), but P^2 is a non-orientable 2-manifold. The easiest way to see this is to think of a theoretical coin heads uppermost near the north pole on S^2 being slid around to near the south pole and then being projected back to near the north pole directly by the antipodal map: it ends up tails uppermost. This corresponds to sliding the coin right

around a track (which must be a Möbius strip) in P^2, since as far as P^2 is concerned the north pole and south pole of S^2 are the *same* point. Since P^2 cannot be realized as a submanifold of R^3 it is impossible to draw a picture of it.

The whole construction generalizes easily to define the n-dimensional *projective n-space* P^n, and there is also a complex version (of real dimension 2n) CP^n derived from C^{n+1} by considering two n-tuples as equivalent when one is a complex scalar multiple of the other, this being the complex analogue of 'lying on the same line through the origin' in R^{n+1}. In fact P^n is non-orientable precisely when n is *even*, and CP^n is always orientable.

3. *The Klein bottle.* A torus can be thought of as a cylindrical pipe with the ends joined up in a natural way. If the ends are joined up in the sense opposite to the natural one (impossible to do physically, but easy enough mathematically) the resulting object is also a 2-manifold, known as the *Klein bottle* (F. Klein, 1849-1925). It is non-orientable: a coin slid from

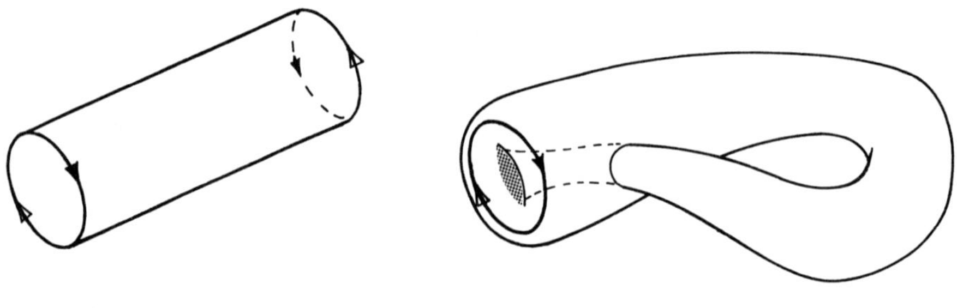

Figure 27

end to end of the cylinder changes from heads to tails on the Klein bottle. It cannot be realized as a submanifold of R^3, although it can be visualized in R^3 by letting it pass through itself somehow, as in Figure 27.

10 After defining smooth manifolds it would be convenient to be able to write down a list of all smooth manifolds - up to diffeomorphism, of course. Since any open set whatsoever in R^n is a smooth manifold, it is obvious that the task is too great. However, we might achieve some results by considering *compact* manifolds. Such progress as has been made can be crudely summarized as follows:

Dimension 1. Every compact connected 1-manifold is diffeomorphic to S^1. (See Guillemin and Pollack [48] or Milnor [83] for proofs.)

Dimension 2. Every compact connected 2-manifold (or *surface*) which is *orientable* is diffeomorphic to one of: *sphere* S^2, *torus* $S^1 \times S^1$, *double torus, treble torus,* ... etc. See Figure 28. The number of 'handles'

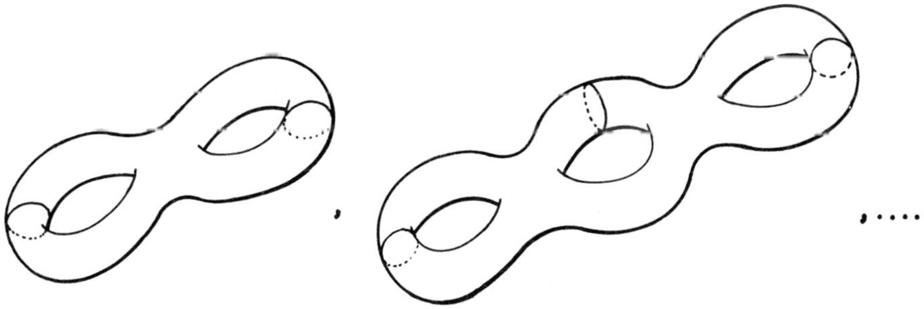

Figure 28

is called the *genus* of the 2-manifold. An analogous classification for *non-orientable* surfaces exists, involving splicing together copies of the projective plane P^2. See Ahlfors and Sario [6], Blackett [16], Massey [77], Wallace [145] for discussions of 2-manifolds.

Dimension 3. No classification is known, or even known to be possible. There are many extremely difficult problems. The most celebrated of these is: does every compact connected simply-connected (every closed loop can be shrunk to a point) 3-manifold have to be diffeomorphic to the 3-sphere S^3 ? This is the *Poincaré Conjecture*.

Dimension ≥ 4. Classification in any reasonable terms was proved impossible by A.A. Markov [72]. However, this does not mean that there are not some very powerful theorems available. For example, Smale [121] proved that the analogue of the Poincaré Conjecture is *true* in all dimensions greater than or equal to 5. There has been a vast amount of very sophisticated algebraic machinery developed for studying the structures and properties of high-dimensional manifolds.

11 As we saw on page 32, the quotient space $\mathbb{R}/\mathbb{Z}$ is homeomorphic to the circle S^1. The usual C^∞ structure on $\mathbb{R}$ provides one for $\mathbb{R}/\mathbb{Z}$, since we can take an atlas for $\mathbb{R}$ whose coordinate charts are open intervals of length < 1 to give an atlas for $\mathbb{R}/\mathbb{Z}$, and then we see that $\mathbb{R}/\mathbb{Z}$ is diffeomorphic to S^1 with its standard C^∞ structure. Similarly, the torus $S^1 \times S^1$ can be regarded as $\mathbb{R} \times \mathbb{R}/\mathbb{Z} \times \mathbb{Z} = \mathbb{R}^2/\mathbb{Z}^2$ by copying the above in each factor independently. More generally, $S^1 \times S^1 \times \ldots \times S^1$ (n times) $= \mathbb{R}^n/\mathbb{Z}^n$. This interpretation is useful for describing explicitly diffeomorphisms or other transformations of the tori, such as the

hyperbolic toral automorphisms in §4.5. Many interesting manifolds (including all *surfaces* as above) can be represented as quotient spaces of euclidean space, although this does not apply to all manifolds. A quotient space of euclidean space or of a manifold will not usually be a manifold.

12 A smooth manifold G which has the algebraic structure of a *group* and for which the operations of multiplication $G \times G \to G : (x,y) \to xy$ and inversion $G \to G : x \mapsto x^{-1}$ are both smooth maps is called a *Lie group* (S. Lie, 1842-1899). Examples are: the circle S^1 (with multiplication $(e^{i\theta}, e^{i\phi}) \mapsto e^{i(\theta+\phi)}$), the torus $S^1 \times S^1$ (with S^1-multiplication taking place in each factor), the group $GL(n;R)$ of invertible $n \times n$ matrices with real entries. This last is an open subset of R^{n^2}. Many Lie groups such as these arise naturally as groups of self-transformations of some geometric object. Conversely, one can start with an abstractly-defined Lie group, and look for objects on which it can act as a group of transformations. There is a particular rich theory of the actions of Lie groups as groups of diffeomorphisms of manifolds. Technically, an *action* of G on a manifold M is a map

$$\alpha : G \to \{\text{all diffeomorphisms } M \to M\}$$

which respects the rules of composition on both sides, i.e.

$$\alpha(y) \cdot \alpha(x) = \alpha(xy)$$

for all x,y in G. (Remember the composition on the left hand side means 'do $\alpha(x)$ then $\alpha(y)$'.) If the map $G \times M \to M$ taking (x,p) to $\alpha(x)(p)$ is a smooth map, then we say that α is a *smooth action*.

For a given point p in M, the set of all points $\{\alpha(x)(p)\}$ as x runs through all of G is called the *orbit* of p under the action α.

It is an immediate consequence of the definition of a group that 'belonging to the same orbit' is an equivalence relation on G, and so M is partitioned into α-orbits. In Chapter 4 we shall be studying the orbit structures of smooth actions of the 1-dimensional Lie group R and the 0-dimensional Lie group Z (in both of which 'multiplication' is usual addition) on compact manifolds.

3.3 THE STRUCTURE OF DIFFERENTIABLE MAPS BETWEEN MANIFOLDS

We have now reached the stage of generalizing the *local* study of maps $f : \mathsf{R}^n \to \mathsf{R}^m$ (i.e. families of real functions of real variables) to the *global* study of maps between differentiable manifolds. There are two main questions to ask:

1. What are the possible geometrical (topological) forms that differentiable manifolds can take?
2. What can differentiable maps between them look like?

Question 1 was discussed briefly in §3.2, [10]. We will now turn attention to Question 2, which is more important in applications since in practice we are likely to know which manifolds we are dealing with but need to know something about maps between them as relating to inputs and outputs of dynamical systems.

To investigate the structure of differentiable maps we first look at local structure and then piece together local information to give global information. As before, we will take *smooth* to mean either C^∞ or C^r for some given $r \geqslant 1$.

(a) *Local structure*

Let M,N be smooth manifolds modelled on R^m, R^n or, more generally, on normed linear spaces E,F. Any smooth map $f : M \to N$ is by definition (§3.2) represented locally by smooth maps between open sets in E and open sets in F, and therefore the theory of the local structure of smooth maps between manifolds is precisely the theory of the local structure of smooth maps in euclidean or more general normed linear spaces.

We have already studied the local structure of such maps to some extent in §§2.8, 2.9 and we have no further results to offer at this stage. However, it is worthwhile looking again at the consequences of the Inverse Function Theorem and thinking of them from a manifold point of view.

Recall that two significant corollaries of the theorem were summarized on page 98 by saying that if $f : R^n \to R^m$ is such that the linear map $Df(x) : R^n \to R^m$ has maximal rank then near x the map f looks locally like Df(x). In particular, when $n \leqslant m$ the local image of f looks like R^n in R^m, and when $n \geqslant m$ the map f looks locally like a linear projection of R^n onto R^m and so the inverse image of each point of R^m near f(x) looks locally like R^{n-m} in R^n. Now the first of these two cases can be described in the language of §3.1 by saying that locally the image of f is an *n-dimensional submanifold of* R^m, and in the second case the local inverse images of points near f(x) are (n-m)-*dimensional submanifolds of* R^n. Of course to tie this down precisely, we have to check that the formal definitions of 'looks like' in the two contexts do correspond to what is required for the formal definition of submanifold in each case, but they do. Thus, remembering the extension of the idea of submanifold from a euclidean space setting to manifolds in general (§3.2 [6]), we can summarize the two corollaries of the Inverse Function

Theorem this time as a theorem about maps between manifolds as follows:

THEOREM

Let $f : N \to M$ *be a smooth map between smooth manifolds of dimensions* n,m *respectively, and let* x *be a point of* N. *Take a local representative* $\bar{f} = \psi f \phi^{-1}$ *for* f *in a neighbourhood of* x, *and suppose that* $D\bar{f}(\phi(x))$ *has maximal rank. Then in the two cases* (i) $n \leqslant m$ *and* (ii) $n \geqslant m$ *we have*

(i) f *takes some neighbourhood of* x *in* N *to an* n*-dimensional smooth submanifold of* M ;

(ii) *the inverse images under* f *of points in some neighbourhood of* $f(x)$ *are locally* (i.e. *near* x) $(n-m)$*-dimensional submanifolds of* N.

Remarks

1. If the maximal rank condition holds for one such local representative $\bar{f}$ then it holds for them all (at the points corresponding to the given x, of course). This follows immediately from the Chain Rule, using the fact that if $A : R^n \to R^m$ is a linear map and $B : R^n \to R^n$, $C : R^m \to R^m$ are isomorphisms then CAB^{-1} has the same rank as A. Arguments of this kind concerning properties of f which are independent of the choice of local representative are becoming quite familiar by now.

2. The case $n = m$ is common to both (i) and (ii). Here the maximal rank condition is that derivative should be an isomorphism; the conclusion is that f is locally a diffeomorphism (the Inverse Function Theorem directly). The local image of f is an open set in M, thus m-dimensional, and locally the inverse images of points near $f(x)$ are single points and thus 0-dimensional.

(b) *Non-singular global structure*

Since the maximal rank property at one point implies that $f : N \to M$ has agreeable local properties near that point, we might expect that if f has the maximal rank property (i.e. using local representatives via charts) *everywhere* than f would have some easily understandable global description. In fact, looking more closely at the above theorem, we could reasonably hope that under the assumption of maximal rank everywhere

(i) if $n \leqslant m$ then fN is a smooth n-submanifold of M

and (ii) if $n \geqslant m$ then inverse images of points are smooth (n-m)-submanifolds of N.

A little thought shows that (i) will not hold as stated, because even though f takes small open pieces of N into submanifolds of M, different pieces on N may collide or otherwise interfere with each other when f is applied and so fN, as a subset of M, may not have the submanifold property. For example, smooth maps $S^1 \to R^2$, $R \to R^2$ as in Figure 29 may well have maximal rank everywhere (in these cases the

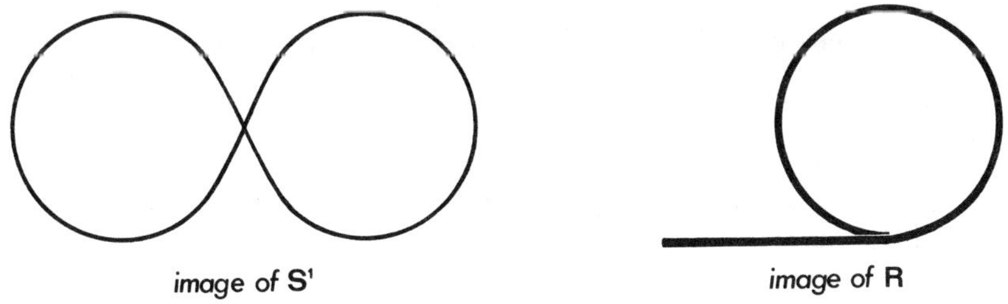

Figure 29

condition amounts to the non-vanishing of the 'velocity' vector $f'(\theta)$) but their images are not submanifolds of R^2 : in (a) there are places - the crossover points - where fS^1 does not look locally like R^1 in R^2, and in (b) the same applies at p even though there is no actual crossover, i.e. f is injective. Nevertheless there will be many instances where fN is a submanifold of M, and when we add the extra assumption that f is a *bijection onto its image*, to rule out examples such as $f : S^1 \to R^2 = C$ taking z to z^2 and thus winding S^1 twice around its image (also S^1), we have the definition of an *embedding*, the most 'nonsingular' kind of map between manifolds $N \to M$ with $n \leqslant m$. Rather than recording the ingredients of this definition directly, we will pause for a moment to see what it means. If we know that fN is a submanifold of M and that $f : N \to M$ satisfies the maximal rank condition, then by taking 'submanifold' charts for fN that exhibit fN in M locally as R^n in R^m (§3.1) we see that the local representatives for $f : N \to fN$ all have non-singular derivative everywhere. Now the Inverse Function Theorem immediately tells us that $f : N \to fN$ is everywhere locally invertible as a smooth map (i.e. locally a diffeomorphism), and since smoothness is a local property this means that the inverse of the bijection $f : N \to fN$ is smooth, or in other words that f is a *diffeomorphism onto its image* fN . Thus we can paraphrase the definition of embedding as follows:

DEFINITION

A smooth map $f : N \to M$ *is an* embedding *if it is a diffeomorphism from* N *to a smooth submanifold of* M.

Thus an embedding of N in M can be regarded as a 'realization' of N as a submanifold of M.

If f satisfies the maximal rank condition (with $n \leq m$) but possibly fails to be an embedding then it is called an *immersion*. Its image is sometimes called an *immersed submanifold* of M, although it may not strictly speaking be a submanifold. From the remarks in (a) we can say that *an immersion is a map which is everywhere locally an embedding*, while it may fail to be an embedding for certain global reasons. Two possible reasons have been illustrated in Figure 29. Here is an important example relating to almost periodic motions in dynamical systems (see §4.1), giving yet another illustration.

EXAMPLE (Irrational flow on a torus)

Let $f : \mathbb{R} \to S^1 \times S^1$ be defined by $f(t) = (\exp i\alpha t, \exp i\beta t)$ where α, β are chosen to be two nonzero real numbers with α/β *irrational*. Then f satisfies the maximal rank condition, since in local angular coordinates on $S^1 \times S^1$ we have $f'(t) = (\alpha,\beta) \neq (0,0)$. Moreover, f is injective, since $f(t) = f(s)$ would imply $\alpha(t-s)$ and $\beta(t-s)$ were both integer multiples of 2π, and so if $t \neq s$ we would have α/β expressible as the quotient of two integers. Thus f is an immersion and is injective, but it is not an embedding since its image $f\mathbb{R}$ is a line winding round and round the torus $S^1 \times S^1$ at rate α in one sense and β in the other and never joining up with itself: it forms a dense subset of the torus and so nowhere looks simply like $\mathbb{R}^1$ in $\mathbb{R}^2$ locally.

Notice that if α/β were rational, then f would still be an immersion but no longer injective. Its image would be (diffeomorphic to) a circle winding p times around the torus one way and q times the other way, where $\alpha/\beta = p/q$ in lowest terms.

The two examples we have seen so far of *injective* immersions that are not embeddings fail to be embeddings because of the way the immersion behaves as points in the domain 'approach infinity': it is only the long-term behaviour that causes the trouble. Now, the topological notion of *compactness* (§1.6) in a sense corresponds to an intuitive idea of not being able to 'approach infinity'. Therefore it is reasonable to expect that every injective immersion of a *compact* manifold may have to be an embedding. This is indeed true.

THEOREM

An injective immersion of a compact manifold is an embedding.

The value of this theorem is that if we are given $f : N \to M$ with $n \leqslant m$, then we only have to check properties of f and N to know that fN will be a submanifold of M: *we do not have to know in advance what* fN *is* to see whether $f : N \to fN$ is a diffeomorphism. This is typical of the kind of simplification to a problem that the use of compactness can give.

Sketch of proof: As mentioned above, the Inverse Function Theorem guarantees that f takes suitably small open neighbourhoods U in N diffeomorphically to their images in M. Once we know that these images are themselves open sets in fN (with its topology induced as a subset of M) we immediately see that $f^{-1} : fN \to N$ is smooth since smoothness is a local property. Therefore the problem is boiled down to proving that f takes open sets in N to open sets in fN, i.e. that f, already a bijection $N \to fN$, is actually a *homeomorphism*. But now we need go no further, for Theorem B in §1.6 tells us that this is true since N is compact by assumption and fN is Hausdorff because M is.

In this discussion of global non-singular behaviour so far we have been concentrating on the case (i) $n \leq m$. What about the other case (ii) $n \geq m$? Here there are no hidden traps. Under the maximal rank assumption, inverse images of points are smooth $(n-m)$-submanifolds of N locally, and therefore also globally since being a submanifold of N is a local property in terms of the topology of N.

In the case $n \geq m$ a map satisfying the maximal rank condition is called a *submersion*, and so we have the following theorem:

THEOREM

If $f : N \to M$ *is a submersion then for every point* p *of* fN *the inverse image* $f^{-1}(p)$ *is an* $(n-m)$-*submanifold of* N.

Remarks

1. There may be points p of M not in the image of f, in which case $f^{-1}(p)$ will be empty.

2. If f is of class C^r then each $f^{-1}(p)$ is in fact a C^r submanifold of N.

3. The simplest examples of submersions (other than diffeomorphisms, which are simultaneously submersions and immersions) are projections from a cartesian product $N = M_1 \times M_2$ onto one or other of the factors. The inverse image of a point p_2 of M_2 under the projection $N \to M_2 : (p_1,p_2) \mapsto p_2$ is the set $M_1 \times \{p_2\}$, which is of course diffeomorphic to M_1. Here $\dim(M_1) = \dim(N) - \dim(M_2)$, as in the theorem.

4. We know from the previous discussion that every submersion has this type of product structure *locally*. That it need not do so *globally* is easily illustrated by the Möbius strip (§3.2 [9]). First remove the boundary of the strip, leaving a non-compact 2-manifold (without boundary), and collapse this manifold down on to a circle going around the strip once. Such a map can be given by $(x,y) \mapsto (x,\frac{1}{2})$ in our formal definition of the Möbius strip as a quotient space obtained from the square $A = [0,1] \times [0,1]$. Locally this looks like a product projection as above, but it is not so globally since the Möbius strip is not homeomorphic to the product $S^1 \times [0,1]$.

This example, showing how a space which is not a cartesian product can nevertheless look everywhere locally like one and even have a globally defined 'projection' (= submersion) onto one 'factor', gives the germ of the topological idea of a *bundle*. We shall meet this again shortly in the context of tangent and cotangent bundles for smooth manifolds.

5. The purpose of removing the boundary of the Möbius strip in the preceding example was to avoid having to define submersions for manifolds with boundary. In fact, with a little care there is no real problem in defining submersions, immersions and embeddings for such manifolds.

6. In general this discussion goes over to infinite-dimensional manifolds, provided the usual precautions are taken about splitting (Remark 1, page 99).

(c) *Global structure with singularities*

The problem of understanding the global structure of smooth maps between manifolds, where singularities are admitted, is no less than the problem of understanding everything about the global properties of families of smooth functions of n variables. Therefore it is not surprizing that our knowledge is rather limited. Nevertheless, there *are* some useful general theorems, and in particular cases very strong results about structure can be obtained. The theory of singularities of smooth maps, and in particular the global theory, is still very young as a branch of mathematics, but it is growing rapidly with the intensive cultivation of many ideas due in large part to Thom in the 1960's.

The simplest non-trivial maps to look at are those for which at least one of the manifolds is R or an interval in R. A map $R \to M$ is thought of as a *path* in M, and although the sets of paths and loops and deformations of loops in M are interesting and important things to study from the point of view of the global topology of M, the actual paths do not lend themselves to much analysis which is of topological significance. After all, a singularity is simply a place on the path where in some (and hence any) local coordinates the 'velocity' vanishes. Therefore we will move immediately on to the more interesting case of maps $M \to R$, i.e. real-valued functions of many variables.

The structure of real-valued functions

Singularities of such functions are called *critical points* and we have already studied some local structure near critical points (using the language of *germs*) in §2.9. We will now consider some global implications.

EXAMPLES

1. Let $M = R^n$ with $f : M \to R$ defined by

$$f(x_1,x_2,\ldots,x_n) = a_1x_1^2 + a_2x_2^2 + \ldots + a_nx_n^2$$

where the a_i's are non-zero constants. Then $Df(x)$ is the linear map

$R^n \to R$ described by the $1 \times n$ matrix

$$(2a_1x_1, 2a_2x_2, \ldots, 2a_nx_n)$$

which fails to have maximal rank (= 1) precisely when $x_1 = x_2 = \ldots = x_n = 0$. Therefore if $k \neq 0$ and k lies in the image of f, the set

$$f^{-1}(k) = \left\{ x \in R^n \;\middle|\; a_1x_1^{\,2} + a_2x_2^{\,2} + \ldots + a_nx_n^{\,2} = k \right\}$$

will be an (n-1)-dimensional smooth submanifold of R^n. In particular, taking each $a_i = 1$ gives that the (n-1)-sphere S^{n-1} is a smooth submanifold of R^n. We already know this (§3.1) but now we have proved it by general theory rather than by considering charts. If all the a_i are positive and $k > 0$ then $f^{-1}(k)$ is a compact manifold diffeomorphic to S^{n-1} (an ellipse when $n = 2$), while if some a_i and a_j have opposite signs then $f^{-1}(k)$ is noncompact (a hyperbola when $n = 2$). Note that $f^{-1}(0)$ is either just one point or is not a submanifold of R^n (a pair of lines when $n = 2$), depending on the a_i's. See Figure 30.

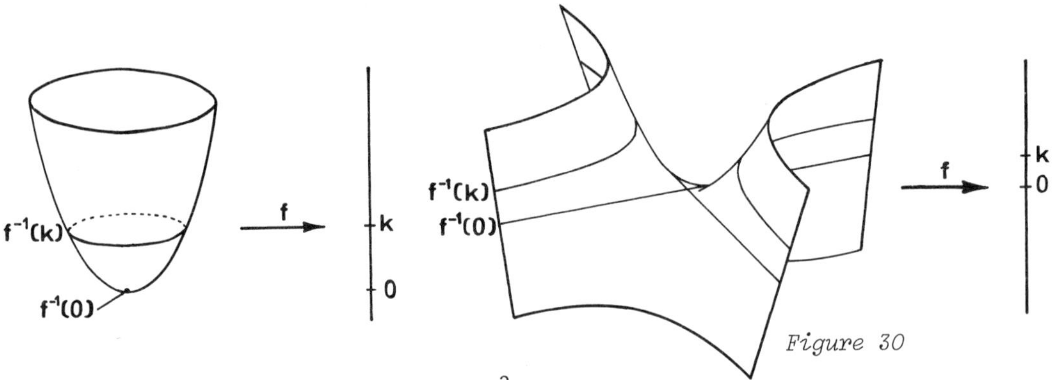

Figure 30

2. Let M be a torus embedded in R^3 in a rather bent way and let $f : M \to R$ be projection onto the x_3-axis (the 'height' function) as sketched in Figure 31. We see that for most points k in fM the inverse image $f^{-1}(k)$ is a 1-dimensional submanifold of M, although not

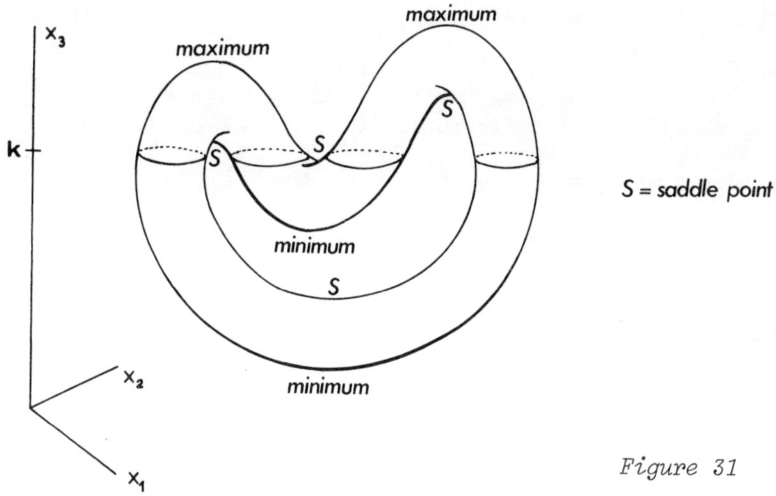

Figure 31

necessarily connected.

From the picture in Example 2 it seems reasonable to expect that in general 'most' points k in the image of f will have inverse images which are $(n-1)$-dimensional (i.e. codimension 1) submanifolds of M. This is indeed the case. It is clearly going to be true when f has only a finite number of critical points, for on removing the critical points the restriction of f to the remaining manifold is always a submersion, but the importance of the result is that it holds for any smooth f whatsoever, no matter how large or complicated the set of critical points may be.

Define a *critical value* of $f : M \to R$ to be any number $k = f(p)$ where p is a critical point of f. Thus critical *points* belong to M, critical *values* to R. Numbers which are not critical values, including those not even in fM, are called *regular values*. Assume that f is C^∞. The theorem that gives us what we hope for is:

THEOREM (Sard's Theorem)

The set of regular values of f *is a dense subset of* R.

Remarks

1. The original theorem (see Sard [111]) is much stronger than this. It relates the *measure* (a non-topological concept) of the set of critical values to the degree of differentiability of f and the dimension of M. Also, it deals with maps from n to m dimensions rather than just real-valued functions: see below.

2. A version of Sard's theorem applicable to infinite-dimensional manifolds has been formulated by Smale [123].

3. The theorem does *not* say that the set of non-critical points is dense in M. For example, the map $R^n \to R$ taking everything to zero has all non-zero k as regular *values*, but has no non-critical *points* at all.

The practical interpretation of the theorem is that if k is such that $f^{-1}(k)$ is non-empty but is not a codimension-1 submanifold of M, then we can find k' as close as we wish to k with the property that $f^{-1}(k')$ is either empty or is genuine codimension-1 submanifold of M.

If we call a codimension-1 submanifold a *hypersurface*, then we can summarize the above discussion by saying that for *'most'* k *in* R *the set of points* x *in* M *satisfying the nonlinear equation*

$$f(x) = k$$

is either empty or is a hypersurface in M.

The advantage of this conclusion is that it allows us to take local coordinate systems consistently and meaningfully everywhere on $f^{-1}(k)$, which we think of as the world in which x can vary subject to the *constraint* $f(x) = k$. We shall use this hypersurface property when considering certain important types of dynamical system (§4.6).

Let us look again at the illustration in Example 2, and suppose that the critical points are given to be non-degenerate. As we saw in §2.9 this definition does not depend on the choice of local coordinate chart. In a neighbourhood of each critical point p the Morse lemma gives us a local model which enables us to see how the submanifold $f^{-1}(k)$ changes or even disappears as k passes through the corresponding critical value $k = f(p)$. Furthermore, $f^{-1}(k)$ does not change at all (up to diffeomorphism) as k moves throughout any interval in R not containing any critical values. Thus from a knowledge only of the critical points of f and their types it seems that we can more or less reconstruct the global geometry of the manifold M. This simple yet profound observation is the basis of *Morse theory*.

Morse theory has been used to great effect by topologists in studying the geometry of high-dimensional manifolds, for example in Smale's proof of the generalized Poincaré conjecture (§3.2, [10]). In practical applications, however, it may often be more useful to work in the opposite direction, using the fact that *the global topological structure of* M *places constraints on the possible numbers of the various types of non-degenerate critical point that a smooth function on* M *can have.*

Let us suppose that M is a *compact* manifold. As remarked in §2.9, each non-degenerate critical point of f is isolated, i.e. lies in a neighbourhood containing no other critical points. If we now suppose also that *all* the critical points are non-degenerate, the compactness of M implies that there can only be a finite number of them (if there were an infinite number they would have to have an accumulation point (§1.6), and since all derivatives of local representatives of f are continuous such an accumulation point would have to be a critical point of f : an impossibility if all critical points are isolated). Let N_i denote the number of critical points of index i (see §2.9). Then the numbers N_i $(0 \leqslant i \leqslant n)$ are related to the topology of M by the so-called *Morse inequalities:*

$$(r < n) \quad N_r - N_{r-1} + N_{r-2} - \ldots \pm N_0 \geqslant B_r - B_{r-1} + B_{r-2} - \ldots \pm B_0$$

$$N_n - N_{n-1} + N_{n-2} - \ldots \pm N_0 = B_n - B_{n-1} + B_{n-2} - \ldots \pm B_0$$

where the B_i are the *Betti numbers* of M (certain integers defined from the topology of M via its *homology groups*: we will give no details here. See e.g. Maunder [79]). The alternating sum

$$B_0 - B_1 + B_2 - \ldots \pm B_n$$

is the *Euler characteristic* of M, usually written $\chi(M)$. The Betti numbers, and hence the Euler characteristic, are *topological invariants*, which means that homeomorphic spaces have the same Betti numbers. These invariants can in principle be computed directly for any manifold or reasonable topological space defined in some fairly concrete way, and much can be discovered about them abstractly using the machinery of homology theory, itself a small part of algebraic topology.

EXAMPLES

1. For the function on the torus in Figure 31 we have $N_0 = N_2 = 2$ and N_1 (the number of saddle-points) = 4, so $\chi(\text{torus}) = 0$. Hence *any* $f : \text{torus} \to R$ with all critical points non-degenerate would satisfy $N_0 - N_1 + N_2 = 0$. It is worth trying several examples to verify this.
2. The n-sphere S^n admits a smooth map $f : S^n \to R$ having one non-degenerate maximum (index n) one non-degenerate minimum (index 0), and no other critical points. For example, let f be the projection of the

'standard' S^n (unit sphere in R^{n+1}) onto one coordinate axis in R^{n+1}. Hence $\chi(S^n) = 1 + (-1)^n$. In particular $\chi(S^2) = 2$, which incidentally gives a proof that S^2 is not homeomorphic to a torus.

3. An orientable surface of genus g (see §3.2, 10) has Euler characteristic 2-2g, as may be seen by counting critical points of a height function based on Figure 28 . Note that this is negative for all cases other than the sphere or the torus.

Maps between arbitrary manifolds

As soon as we take both manifolds to have dimension greater than 1 the global problems become much more subtle and complicated. However, one of the main tools - Sard's Theorem - is still available. Following a path parallel to that we took for functions $f : M \to R$, we define *singular points* of $f : M \to N$ to be points p where the maximal rank condition fails (in some and therefore any local coordinates), *singular values* to be points q of N which are the images of singularities under f, and *regular values* to be points in N which are not singular values (including points which are not in fM). Then, assuming f to be C^ω, Sard's Theorem says that *the set of regular values is a dense subset of* N.

Again exploiting the corollaries of the Inverse Function Theorem as in the first part of this section, we see that when dim N $\leqslant$ dim M the inverse image of each regular value is either empty or is a smooth submanifold of M of codimension equal to dim N. Thus the practical interpretation is that if $f^{-1}(q)$ is non-empty but is not a submanifold of M of codimension equal to dim N, then there exist points q' arbitrarily close to q in N for which either the submanifold condition *will* be true or $f^{-1}(q')$ will be empty. Much as before, we can summarize this by saying that *'usually' the set of solutions of* n *non-linear equations in* m *variables, where* $n \leq m$, *is either empty or is a smooth manifold of dimension* $m-n$.

Given some notion of 'nondegeneracy' for singularities in arbitrary dimensions, we would aim to mimic the Morse inequalities by relating the number and types (whatever they may be) of nondegenerate singularities of maps $f : M \to N$ to the topological structures of M and N. This is a programme which has hardly begun, since it is only relatively recently that suitable ideas of 'nondegeneracy' in this wider context have been evolved, and in any case the difficulties seem formidable. It is likely, though, that we shall see exciting developments in these directions in the next few years.

3.4 TANGENT BUNDLES AND TANGENT MAPS

In the previous discussion of the structure of smooth maps the maximal rank condition played an important part. This is a condition on the derivative of a local representative of a map, and carries with it the remark that it does not depend on the choice of local representative. Similarly, the definitions of *critical point, non-degenerate, singular point* are in terms of the derivatives of a local representative, with the accompanying assurance about independence of choice of representative. It is cumbersome to have to refer to local representatives when dealing with properties independent of them, but how else can one work with derivatives? The answer is: by means of the *tangent bundle,* a concept which will enable us to deal with derivatives of smooth maps without mentioning local representatives explicitly.

If we think of some evolving dynamical system as being represented by a point moving on a manifold, then it is likely that we will be interested not only in the position but also in some sense the *velocity* of the point. The intuitive idea of the tangent bundle of M is that it is the space of all *positions and velocities* of points moving on M. The subtlety of its construction stems from the need for an intrinsic definition, using information only about M and not about any embedding of M in some euclidean or other linear space (in which case the definition of velocity vectors would present no problem).

To formulate the definition let us go back to a local setting. Take E, F to be Banach spaces (e.g. euclidean spaces R^n), and let U, V be open sets in E, F respectively. The basis of our definition of tangent bundle will be the following observation: if $f : U \to V$ is differentiable at p in U, then *the derivative* $Df(p) : E \to F$ *is characterized entirely by the effect of* f *on smooth paths in* U *based at* p. Some explanation is needed. By a smooth path in U based at p we mean a smooth map

$$c : J \to U$$

where J is some open interval (a,b) with $a < 0 < b$ and $c(0) = p$. If $f : U \to V$ is a smooth map, then by the Chain Rule we see that the composition

$$f \circ c : J \to V$$

is a smooth path in V based at $c(p)$ and that

$$D(f \circ c)(0) = Df(c(0)) \cdot Dc(0) : R \to F,$$

i.e. $$(f \circ c)'(0) = Df(p) \cdot c'(0) \quad . \qquad (*)$$

Thus *the derivative* $Df(p)$ *takes the tangent to the path* c *at* p *to the tangent to the path* $f\cdot c$ *at* $f(p)$. See Figure 32.

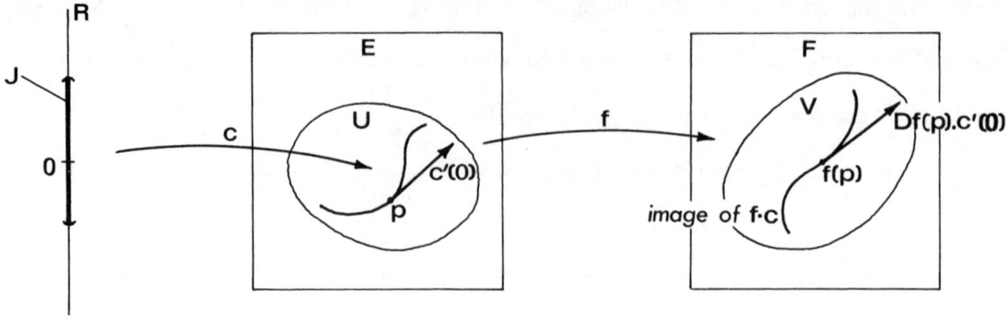

Figure 32

Now *every* vector v in E is the tangent to some path in U based at p (e.g. $c : t \mapsto p + tv$, for $|t|$ small enough to ensure $c(t)$ lies in U) and so the property above describes $Df(p)$ completely.

Given two smooth paths c_1, c_2 based at p, define them to be *tangent* there if

$$(c_1(t) - c_2(t))/t \to 0 \text{ (in E)} \quad \text{as} \quad t \to 0$$

(the usual definition of 'first order contact'). Then tangency is an equivalence relation on the set of all paths in U based at p, and we call the equivalence classes *tangency classes* at p. This is the formal way of capturing the idea of *velocities* at p.

Denote the tangency class of c by $[c]$. Then the map

$$\sigma_p : \{\text{tangency classes at } p\} \to E$$

defined by

$$\sigma_p[c] = c'(0)$$

makes sense ($[c_1] = [c_2]$ implies $c_1'(0) = c_2'(0)$) and is injective ($c_1'(0) = c_2'(0)$ implies $[c_1] = [c_2]$) and we have already noted above that it is surjective (for any v in E there exists c with $c'(0) = v$).

Therefore we have a natural bijection between the linear space E and the set T_pU of tangency classes of smooth paths in U based at p. This means that we can regard T_pU as a normed linear space isomorphic to E (just lift the structure of E over to T_pU via the bijection), and similarly think of $T_{f(p)}V$ as an isomorphic copy of F. This linear space T_pU is called the *tangent space* to U at the point p. In this new guise the linear map $Df(p) : E \to F$ becomes the linear map

$$T_pf : T_pU \to T_{f(p)}V$$

which takes each $[c]$ in T_pU to $[f \circ c]$ in $T_{f(p)}V$.

Define the *tangent bundle* TU of U to be the union of all the linear spaces T_pU as p runs through U. The isomorphisms $\sigma_p : T_pU \to E$ fit together to give a bijection

$$\sigma : TU \to U \times E$$

defined explicitly by $\sigma([c]) = (p,c'(0)) = (c(0), c'(0))$ for every smooth path c based at p. We then assign to TU that unique topology which makes σ into a homeomorphism, i.e. W is *open* in TU precisely when σW is open in $U \times E$ with its product topology.

Any smooth map $f : U \to V$ induces a *tangent map*

$$Tf : TU \to TV$$

defined as T_pf on each linear space T_pU. Thus Tf represents the effect of the derivative of f simultaneously all over U. If we regard TU, TV as $U \times E$, $V \times F$ via σ and its analogue for V, then Tf is the map

$$U \times E \to V \times F : (p,h) \mapsto (f(p), Df(p).h) \quad . \qquad (**)$$

From this expression we see that if f is of class C^r then Tf is of class C^{r-1} since Df is of class C^{r-1} (cf. page 80).

Using this notation the Chain Rule has a very clean form. If we have smooth maps f, g whose composition $g \cdot f$ makes sense then $T(g \cdot f)$ takes (p,h) to

$$(g \cdot f(p),\ D(g \cdot f)(p).h) = (g(f(p)),\ Dg(f(p)) \cdot Df(p).h)$$
$$= Tg(Tf(p,h))$$

so in fact the Chain Rule becomes

$$T(g \cdot f) = Tg \cdot Tf.$$

This is easy to see from the tangent space definition, since $T(g \cdot f)([c]) = [gfc] = Tg([fc]) = Tg \cdot Tf([c])$ for any $[c]$ in TU.

From the Chain Rule it follows immediately that if f is a diffeomorphism then so is Tf, since $T(id_U) = id_{TU}$ (similarly for V) and so $T(f^{-1})$ is an inverse for Tf.

The idea of tangency classes is the one we use in order to put derivatives into a manifold context. If M is any smooth manifold modelled on E and p is a point on M, define a *smooth path based at* p to be a smooth map

$$c : J \to M$$

where $J = (a,b)$ with $a < 0 < b$ and $c(0) = p$, and define two such to be *tangent* if their local representatives with respect to some chart around p are tangent in the sense already defined above for maps between open sets in normed linear spaces. *This does not depend on the choice of chart*, as is easily seen by applying (*) in the case when f is an overlap map $\chi = \phi_\alpha \phi_\beta^{-1}$ between two charts, and remembering (page 69) that since χ is a local diffeomorphism it follows that $D\chi(\phi_\beta(p))$ is a linear isomorphism $E \to E$. This tangency is a coordinate-independent property of the paths themselves, and we can define the *tangency class* $[c]$ of c in this manifold setting.

DEFINITION

The tangent space *to* M *at* p *is the set of tangency classes of smooth paths in* M *based at* p. *It will be denoted by* T_pM. *Elements of* T_pM *are called* tangent vectors *to* M *at* p.

A choice of chart (U,ϕ) around p gives us a bijection between T_pM and E by letting $[c]$ correspond to $(\phi c)'(0)$ (note that $c'(0)$ itself would be meaningless), or in other words by applying the bijection σ to the local representative c. If we choose a different chart we may obtain a different bijection between T_pM and E. However, the important thing to notice is that if we go from E to T_pM by the bijection corresponding to one chart, and then from T_pM to E by the bijection corresponding to another chart, the net result is a *linear isomorphism* $E \to E$. This follows again from (*) above. Therefore, *no matter which chart we choose, the linear structure that* T_pM *acquires via the bijection with* E *will always be the same one*. In more concrete terms, this means that given $[c_1], [c_2]$ in T_pM we can define $[c_1] + [c_2]$ in T_pM by choosing a chart (U,ϕ), going over to E, computing $(\phi c_1)'(0) + (\phi c_2)'(0)$, and going back to T_pM: the net result will not depend on the choice of (U,ϕ). The same applies to scalar multiples $\lambda[c]$, and the rules for a linear space (§2.2) work for T_pM because they do for E.

If we have a smooth map $f : M \to N$, where N is a manifold modelled on F, then by taking local representatives we see that if $[c_1] = [c_2]$ in T_pM then $[f \cdot c_1] = [f \cdot c_2]$ in $T_{f(p)}N$. This means that f determines a well-defined map

$$T_pf : T_pM \to T_{f(p)}N$$

$$[c] \mapsto [f \cdot c].$$

In terms of a local representative $\overline{f} = \psi f \phi^{-1}$ this map $T_p f$ is none other than $D\overline{f}(\phi(p)) : E \to F$. Since this is linear, the map $T_p f : T_p M \to T_{f(p)} N$ is a *linear map*. It is called the *tangent map* of f at p. We think of it as taking velocities of points p of M to the corresponding velocities of points f(p) on N.

To summarize: the tangent space $T_p M$ is a linear space isomorphic to E; constructing an explicit isomorphism requires choosing a chart around p, but the linear structure induced on $T_p M$ does not depend on the chart. A smooth map $f : M \to N$ induces a linear tangent map $T_p f : T_p M \to T_{f(p)} N$ by $[c] \mapsto [f \circ c]$, which in local representation is simply the derivative.

Note that we have 'forgotten' the norm structure on E, and we are not thinking of any particular norm for $T_p M$ at the moment. We shall come back to this point shortly.

So far we have discussed only what happens at one point on the manifold. The next task is to put together in some coherent way the tangent spaces and tangent maps over the whole manifold M.

DEFINITION

The tangent bundle *of* M, *denoted by* TM, *is the union of the tangent spaces at all the points of* M.

Equivalently, TM is the set of all tangent vectors everywhere on M. As it stands, this definition gives TM simply as a set, with no further structure than the linear structure of each $T_p M$. However, TM can quickly be made not only into a topological space but into a smooth manifold, and it then turns out that Tf is a smooth (C^{r-1} if f is C^r) map from TM to TN.

Topology and smooth structure for TM

Let us first take the local case, when M is an open subset U of E. Here we have already seen that there is a natural bijection

$$\sigma : TU \to U \times E$$

defined by taking T_pU to $\{p\} \times E$ via σ_p for each p in U, where we think of $\{p\} \times E$ as just a copy of E 'labelled' by p. If we are on an arbitrary C^r manifold M we can take a chart $\phi : U \to U'$ and construct a bijection τ between TU (thinking of U itself as a manifold) and $U' \times E$ by taking each $[c]$ in each T_pU to the point $(\phi(p),(\phi c)'(0))$ in $U' \times E$ – i.e. we identify TU with TU' by letting each path c in U correspond to ϕc in U', and then apply $\sigma : TU' \to U' \times E$. Again, we already have a product topology defined for $U' \times E$, and so we lift this back via τ to a topology for TU. Now this construction depends on the choice of chart $\phi : U \to U'$, but the *topology obtained for* TU *does not depend on the choice of* ϕ. This is because any overlap map $\chi = \phi_\alpha \phi_\beta^{-1}$ is a C^r diffeomorphism, and so as we saw above $T\chi$ is a C^{r-1} diffeomorphism (if $r > 1$) and is in any case a homeomorphism. But it follows directly from (*) that $T\chi$ is in fact $\tau_\alpha \tau_\beta^{-1}$ and, so $\tau_\beta W$ is open if and only if $\tau_\alpha W = \tau_\alpha \tau_\beta^{-1}(\tau_\beta W)$ is open, showing that the topology for TU is independent of ϕ.

For each chart $\phi : U \to U'$ we now have a well-defined topology on TU, independent of ϕ. To get from this a topology on the whole of TM we define a set V in TM to be an *open set* precisely when *the intersection of* V *with every* TU *is open in* TU. It is routine to check that this does satisfy the rules for a topology (§1.3). In view of the overlap condition discussed above, it is not necessary (and of course theoretically impossible) to check this intersection properly for *all* charts on M : it suffices to check it for one atlas.

By means of the bijections τ we have gained more than a topology for TM. Each τ is a homeomorphism (by definition) from an open set TU in TM to an open set $U' \times E$ in $E \times E$, and so TM is a manifold modelled on $E \times E$. Furthermore, we observed above that the overlap maps $\tau_\alpha \tau_\beta^{-1}$ are C^{r-1} diffeomorphisms if the overlap maps $\phi_\alpha \phi_\beta^{-1}$ for M are C^r, and so a C^r structure on M automatically provides a C^{r-1} atlas and hence a C^{r-1} structure for TM.

Remarks

1. If M is C^∞ or C^ω then the same holds for TM.
2. Technically, we should check that TM is Hausdorff and has an atlas with at most a countably infinite number of charts (see page 122). The second condition clearly holds since we constructed an atlas for TM from any given atlas for M; the first condition is easily verified by considering separately the cases of (a) two points in the same tangent space, (b) two points in different tangent spaces.
3. If M has dimension n then TM has dimension 2n.
4. If $\dim M > 0$ then TM is never compact. This is essentially because each T_pM is non-compact, being homeomorphic to E. (A compact set in E would have to be bounded: cf. Theorem C in §1.6.)
5. There is a natural map $\pi : TM \to M$, called the *projection* map, taking T_pM to p for each p in M, i.e. π takes all tangent vectors at p to the point p itself. Thus $\pi^{-1}(p) = T_pM$. With π in mind we think of the tangent bundle TM as a collection of tangent spaces lying 'above' M, and then T_pM is called the *fibre* of the fundle over p. See Figure 33.

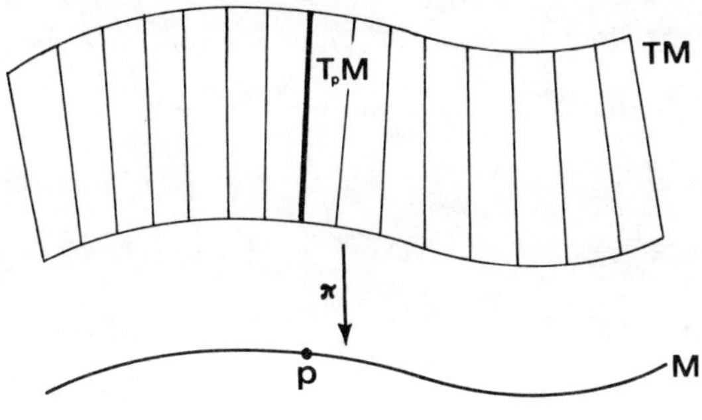

Figure 33

The projection π is a smooth map (C^{r-1} if M is C^r) since it has local representatives of the form $\phi\pi\tau^{-1}$ taking (q,h) in $U' \times E$ to q in U'. This has maximal rank everywhere (its derivative is surjective, and splits in the infinite-dimensional case) and so π is a *submersion* $TM \to M$.

EXAMPLES of tangent bundles

1. For an open set U in E the topology on TU was defined just in order to make the bijection $\sigma : TU \to U \times E$ into a homeomorphism. For this reason we usually write $TU = U \times E$. In effect this says that we are using the same linear space E in which to represent the tangent vectors at all points in U simultaneously. This is useful for calculations, but confusing conceptually, since U itself lies in (one copy of) E and so tangent vectors lying in (the other copy of) E might be mistaken for elements of U. It is important to try to avoid this confusion.

2. If S is a smooth k-dimensional submanifold of M then by taking the right charts for M at points on S we see that TS is a smooth 2k-dimensional submanifold of TM. For each p in S the tangent space T_pS is a k-dimensional linear subspace of T_pM.

3. As an example of the above, consider S^1 in R^2. A path c in S^1 can be written in polar coordinates $c(t) = (r(t),\theta(t))$ where of course $r(t) = 1$ since c is in S^1. The map $\alpha : TS^1 \to S^1 \times R$ taking $[c]$ to $(c(0),\theta'(0))$ for each $[c]$ in TS^1 is clearly a bijection, and by taking charts for S^1 it is easy to check that α is a diffeomorphism. Furthermore α takes each T_pS^1 to R (the second factor) by a linear isomorphism, so we usually write simply $TS^1 = S^1 \times R$. The bundle projection $\pi : TS^1 \to S^1$ is just projection on the first factor.

4. $T(M \times N)$ can be regarded as $TM \times TN$ by working with the first and second 'coordinates' separately.

5. Since TM is itself a smooth manifold, it has its own tangent bundle $T(TM)$ or T^2M of dimension $4n$ if M has dimension n. Similarly we define $T^3M = T(T^2M)$, and so on. If M is of class C^r then TM is C^{r-1}, and so the process stops at T^rM which is a topological manifold but has no given differentiable structure and so no tangent bundle. Locally T^2M looks like

$$T(U \times E) = (U \times E) \times (E \times E)$$

since $U \times E$ is an open subset of the linear space $E \times E$.

6. From Examples 3 and 4 it follows that

$$\begin{aligned} T(S^1 \times S^1) &= (S^1 \times R) \times (S^1 \times R) \\ &= (S^1 \times S^1) \times R^2 \end{aligned}$$

on rearranging the factors. This expresses the fact that angular coordinates θ,ϕ on the two S^1 factors allow tangents to paths on $S^1 \times S^1$ to be described everywhere by the pair of numbers $(\theta'(t),\phi'(t))$ in R^2.

7. The two-sphere S^2 does not have a system of angular coordinates (the Euler angles give charts that do not work at the north and south poles),

and this is related to the fact that TS^2 cannot be identified with $S^2 \times R^2$. If it could be, then we would be able immediately to assign a non-zero tangent vector to each point of S^2 in a way that varied continuously over S^2 (take a fixed vector v in the second factor R^2). The celebrated Hairy Ball Theorem in topology asserts that this is impossible. We have here more than a mere topological curiosity, for we saw on page 131 that the manifold that we now recognize as TS^2 is the natural setting for the study of the motion of a spherical pendulum.

In fact $TS^n = S^n \times R^n$ only in the cases $n = 1,3$ or 7 (or 0). This is a deep result, first established by J.F. Adams [3] in 1962. It is closely connected with the existence of complex numbers (dim 2), quaternions (dim 4), Cayley numbers (dim 8) but nothing analogous in higher dimensions. The proof involves much elaborate machinery from algebraic topology.

The question of whether or not TM is the same as $M \times R^n$ is very important both on purely geometric grounds and from the point of view of applications. If $TM = M \times R^n$ then velocities of points moving on M (= tangents to paths) can all be represented in one n-dimensional coordinate space R^n, despite the fact that M itself may need several coordinate charts. However, if $TM \neq M \times R^n$ then we need separate coordinate charts for velocities as well. Of course if M is embedded as a submanifold in some R^m then TM is a submanifold of TR^m (Ex.2) which is $R^m \times R^m$ (Ex.1), and velocities on M can be measured in the second factor R^m. However, this introduces a lot of redundant extrinsic information about the way M lies in R^m, and is not helpful to an understanding of intrinsic processes on M itself.

A tangent bundle TM which is really a product $M \times R^n$ is called *trivial*. The precise definition, valid also in infinite dimensions, is the following:

DEFINITION

If M *is a smooth manifold modelled on* E *then the tangent bundle* TM *is* trivial *if there exists a diffeomorphism*

$$TM \to M \times E$$

which takes each tangent space T_pM *by a linear isomorphism to the copy* $\{p\} \times E$ *of* E *in* $M \times E$. *A manifold whose tangent bundle is trivial is called* parallelizable.

Tangent maps for manifolds

By analogy with the local case (open set U in E) we define the *tangent map* $Tf : TM \to TN$ of a smooth map $f : M \to N$ to be the map which is the linear map T_pf on each tangent space T_pM. While f takes points on M to points on N, the map Tf takes velocities on M to velocities on N. If we take suitable charts $\phi : U \to U'$ and $\psi : V \to V'$ to look at a local representative $\psi f \phi^{-1}$ for f we see that $T(\psi f \phi^{-1})$ is a local representative for Tf. Therefore Tf is of class C^{r-1} if f is of class C^r.

There is a simple *global Chain Rule* for maps between manifolds: given $f : M \to N$ and $g : N \to P$ we have $g \cdot f : M \to P$ and since $[g \cdot f \cdot c] = Tg \circ Tf([c])$ we see as before that

$$T(g \cdot f) = Tg \cdot Tf.$$

Again, $T(id_M) = id_{TM}$ and it follows by the same formal argument as in the local case that if $f : M \to N$ is a diffeomorphism (C^r) then so is

$Tf\ (C^{r-1})$.

Since Tf takes the tangent space of M at p to the tangent space of N at $f(p)$, we could write

$$\pi_N \circ Tf = f \circ \pi_M : TM \to N$$

where π_M denotes the projection $TM \to M$, and similarly for π_N. This is often conveniently represented by saying that the diagram

$$\begin{array}{ccc} TM & \xrightarrow{Tf} & TN \\ {\scriptstyle \pi_M}\downarrow & & \downarrow{\scriptstyle \pi_N} \\ M & \xrightarrow{f} & N \end{array}$$

commutes, meaning that you arrive at the same result whichever route you follow from top left to bottom right. This diagram conveys the idea of $Tf : TM \to TN$ being a map *over* f, carrying on top of f information about the derivative of f.

EXAMPLES of tangent maps

1. It is worth emphasizing again that when U, V are open sets in E, F and $f : U \to V$ then $Tf : TU \to TV$ is the map

$$U \times E \to V \times F : (p,h) \mapsto (f(p), Df(p)h).$$

This is the form of all local representatives of $Tf : TM \to TN$.
Remember that Df makes no sense in the manifold case, although Tf does. To carry out calculations (such as checking the maximal rank condition) we work with local representatives and D, but for general theorems we use T.

2. Define $f : S^1 \to S^1$ by $f(z) = z^q$ where z is a complex number $e^{i\theta}$ and q is an integer. Then (Example 3 above) $TS^1 = S^1 \times R$ and here $Tf : S^1 \times R \to S^1 \times R$ is given by

$$Tf(z,h) = (z^q, qh).$$

This is because $f(e^{i\theta(t)}) = e^{qi\theta(t)}$ and so f multiplies velocities of paths on S^1 by the factor q.

Transversality

Since the tangent space to M at p is a local linear version of M near p, the geometry of configurations of linear spaces and subspaces can be used to model local geometry of configurations of manifolds and submanifolds. The development of this simple observation into a theory of *transversality of intersection* by Thom (see §4.7) has been of profound importance in many aspects of differential topology.

Two lines in R^3 will in general not meet; two planes will in general meet in a line, and a line and a plane will meet in a point. The geometry extends easily to R^n: two affine subspaces K,L of dimensions k,ℓ will in general not meet unless $k + \ell \geqslant n$, and if the dimension condition is satisfied then K and L will in general meet in such a way that $\dim K \cap L = k + \ell - n$ and $K + L = R^n$: this is easy to believe from considering low-dimensional cases that can be visualized. Applying this to tangent spaces of manifolds, we say that two submanifolds K,L of a finite-dimensional manifold M intersect *transversally* if for every point p on $K \cap L$ we have

$$T_pK + T_pL = T_pM.$$

It is then not difficult to prove using the Inverse Function Theorem that $K \cap L$ is a $(k + \ell - n)$-dimensional submanifold of M locally near p. If $k + \ell = n$ then up to local diffeomorphism M is expressible as the product $K \times L$ near p. If $k + \ell < n$ the transversality condition cannot be satisfied for any p in $K \cap L$ so in this case we take transversality to mean $K \cap L$ is empty.

The idea becomes more useful if instead of two submanifolds we take one submanifold together with the image of another manifold under a smooth map into M. If S is a submanifold of M and $f : N \to M$ is a smooth map we look at points p in N and say that f is *transversal* to S at p if either f(p) does not lie in S or, if it does, then

$$T_p f(T_p N) + T_{f(p)} S = T_{f(p)} M.$$

See Figure 34. It follows from the Inverse Function Theorem that locally

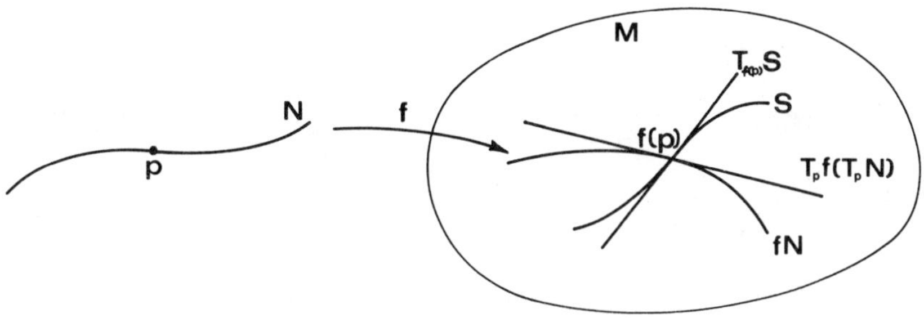

Figure 34

near p the inverse image $f^{-1}(S)$ is (if not empty) a smooth submanifold of N of codimension equal to that of S in M. We have already met this idea before in the case when S is one point: then transversality means surjectivity of $T_p f$ and f is locally a submersion (cf. page 153). The map version of transversality becomes the submanifold version when f is an embedding, and is very similar when f is an injective immersion. We shall use it in this context when studying dynamical systems in §4.4.

Transversality makes sense in infinite dimensions, although to use any of the theory we have as usual to include a proviso about splitting. We might suppose that $T_{f(p)} M$ splits into a direct sum $T_{f(p)} S \oplus F$, say, and that $(T_p f)^{-1}(T_{f(p)} S)$ is a factor in a splitting of $T_p N$, although sometimes we can assume less than this. Transversality theory in function spaces is a very powerful tool in differential topology and global analysis. See Abraham and Robbin [2].

Returning to the original ideas of 'general position' in R^n, we hope that in some sense a map $f : N \to M$ would in general be transversal to a given submanifold S of M. This is the essence of Thom's Transversality Theorem (page 259). However, we cannot treat this here since we have not yet discussed a definition for 'in general'. We shall return to these questions in Chapter 4.

Riemannian structures

As remarked on page 168, the norm on E is forgotten when TM is modelled locally as $U \times E$. Once the tangent bundle has been constructed, however, it becomes useful to have a measure of length in each T_pM not only so that paths in M have speed (a number) as well as velocity (a tangent vector) but also in order to measure global contracting/expanding behaviour of dynamical systems: see §4.5. If E is a Hilbert space H (page 52), such as R^n for example, then it would be useful to be able to equip each T_pM with an inner product $\langle\cdot,\cdot\rangle_p$ which varies continuously or, better, smoothly with respect to p. By this we mean that $\langle X(p),Y(p)\rangle_p$ should be a smooth function of p for all smooth vector fields X,Y defined in a neighbourhood of p. In fact it is always possible to do this, although certainly not in any unique way. The family of inner products is called a *Riemannian structure* (B. Riemann, 1826-1866); it is technically a section of a bundle of covariant tensors (see §4.6). Given a Riemannian structure on M we have on each T_pM a norm $||\cdot||_p$ defined (as on H) by $||v||_p = \langle v,v\rangle_p^{\frac{1}{2}}$ for $v \in T_pM$.

With this norm we can construct a *metric* on M, using the obvious intuitive idea of the 'shortest distance between two points'. Given a C^1 path $\gamma : [0,1] \to M$ with $\gamma(0) = p$, $\gamma(1) = q$ we define the distance from p to q along γ as

$$d_\gamma(p,q) = \int_0^1 ||\gamma'(t)||_{\gamma(t)} dt.$$

Then taking $d(p,q) = \inf_\gamma d_\gamma(p,q)$ gives a metric on M. For this reason the Riemannian structure is itself often called a Riemannian metric.

3.5 VECTOR FIELDS AND DIFFERENTIAL EQUATIONS

Consider a system S of first order autonomous (i.e. no explicit t on the right hand side) ordinary differential equations

$$\begin{aligned}
\dot{x}_1 &= X_1(x_1,x_2,\ldots,x_n)\\
\dot{x}_2 &= X_2(x_1,x_2,\ldots,x_n)\\
&\vdots\\
\dot{x}_n &= X_n(x_1,x_2,\ldots,x_n)
\end{aligned}$$

where $\dot{}$ denotes $\frac{d}{dt}$. We can write the system more economically as

$$\dot{x} = X(x)$$

where x lies in some open subset U of R^n and X is a map from U to R^n, which we do not assume to have any special properties at this stage. It is convenient, and in keeping with our general aim, to regard t here as a measure of *time*. The set U is called the *phase space* of S.

What does it mean to *solve* the equations? First we have to postulate some initial conditions such as

$$x = p \text{ when } t = 0$$

and then a *solution* means a path $c : J \to U$ satisfying

$$c(0) = p$$

and

$$\dot{c}(t) = X(c(t))$$

for all t in the interval J, where $\dot{c}(t)$ is what in §2.4 we called $c'(t) = Dc(t).1$.

We think of X as prescribing a vector $X(x)$ at each point x in U, and we want a path c based at p with the property that the tangent at a typical point $c(t)$, i.e. the velocity of that point, is precisely the vector $X(c(t))$. We should therefore think of $X(x)$ not as an element of R^n in which x lies, but as an element of the tangent space T_xU. Then X becomes a map

$$X : U \to TU = U \times R^n$$

with $X(x)$ an element of T_xU for every x in U, or in other words X satisfies

$$\pi \circ X = id_U$$

where $\pi : TU \to U$ is the natural projection map. Such a map is called a *vector field* on U. Formally, a *solution* c then satisfies $X(c(t)) = (c(t), \dot{c}(t))$, where the seemingly redundant $c(t)$ in the first factor is there to keep track of the particular tangent space in which $X(c(t))$ lies. Since by definition $\dot{c}(t) = Dc(t).1$ we have in tangent map notation

$$X(c(t)) = (c(t), Dc(t).1) = Tc(t,1) \quad \text{or} \quad T_t c(1).$$

These definitions, in which we have disentangled the space in which $X(x)$ lies from that in which x itself lies, generalize immediately to manifolds.

DEFINITION

A vector field *on a smooth manifold* M *is a map* $X : M \to TM$ *which satisfies* $\pi \circ X = id_M$, *where* π *is the natural projection* $TM \to M$.

Thus $X(p)$ lies in $\pi^{-1}(p) = T_pM$ for every point p on M, i.e. X assigns a tangent vector (velocity) at p to each point p. Such a map is called a *section* of the bundle TM. See Figure 35.

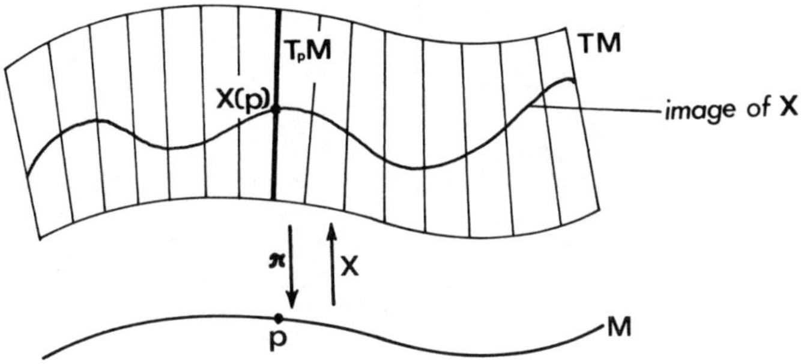

Figure 35

The notion of a *solution* to the system S now becomes:

DEFINITION

Let X *be a vector field on* M. *A* solution curve *to* X, *based at* p *on* M, *is a path* $c : J \to M$ (*where* J *is some open interval* (a,b) *with* $a < 0 < b$) *satisfying* $c(0) = p$ *and*

$$T_tc(1) = X(c(t))$$

for all t *in* J. (*Here* $TJ = J \times R$ *and so* T_tc *is a map* $R \to TM$.)

Thus a vector field on a manifold is the global version of a first order autonomous system of n ordinary differential equations on R^n. Observe that the definition works equally well for smooth manifolds modelled on an infinite dimensional normed linear space E.

Higher order equations

There is a standard trick for converting an nth order equation in one variable into a system of n first order equations in n variables. Assuming the nth order equation is of the form

$$\frac{d^n x}{dt^n} = F\left(x, \frac{dx}{dt}, \ldots, \frac{d^{n-1}x}{dt^{n-1}}\right)$$

we write x_1 for x and then $x_2 = \dot{x}_1$, $x_3 = \dot{x}_2, \ldots,$ and finally

$$\dot{x}_n = F(x_1, x_2, \ldots, x_n) \quad ,$$

giving us a first order system on R^n. The subset of R^n on which all this is defined (generally an open subset) is the *phase space* for the original equation.

In the case $n = 2$ we have

$$\left.\begin{aligned} \dot{x}_1 &= x_2 \\ \dot{x}_2 &= F(x_1, x_2) \end{aligned}\right\}$$

and so from a second order equation on an open interval U in R we obtain a first order equation (i.e. a vector field) on $U \times \mathsf{R} = TU$. Not every vector field X on TU comes from a second order equation on U : it is easy to see that a necessary and sufficient condition on X for this is that the x_1-component of $X(x)$ in $T_x U = \mathsf{R}^2$ should be equal to x_2. In tangent bundle language this says

$$T\pi \cdot X = id_{TU}$$

where $\pi : TU \to U$ is the usual projection, $T\pi : T^2U \to TU$ is its tangent map, and $X : TU \to T^2U$ is the vector field on TU. This generalizes verbatim to manifolds:

DEFINITION

A second order autonomous ordinary differential equation on M *is a vector field* $X : TM \to T^2M$ *on* TM *which satisfies*

$$T\pi \cdot X = id_{TM}$$

where $\pi : TM \to M$ *is the natural projection (so we have* $T\pi : T^2M \to TM$).

Non-autonomous equations

A system of first order equations which contains t explicitly on the right hand side can be regarded as a *family* X_t *of vector fields varying with* t. This goes over directly to vector fields on manifolds. By introducing another variable $u = t$ we can interpret the family of vector fields X_t as just one vector field $\overline{X}$ on the product manifold $M \times R$. At a point (x,u) of $M \times R$ the tangent space to $M \times R$ is $T_xM \times T_uR = T_xM \times R$ and the component of $\overline{X}(x,u)$ in the second factor is 1 since $\frac{du}{dt} = 1$. This technique is particularly useful when X_t is periodic (of period 1, say) in t : in this case $\overline{X}$ projects to a vector field on $M \times (R/Z) = M \times S^1$ giving rise to a flow with a cross-section: see §4.1.

Local existence and uniqueness of solutions

It is a standard and fundamental theorem that provided X is sufficiently well-behaved local solutions to X *exist*, are *essentially unique* and furthermore if X is smooth they *vary smoothly with respect to the initial conditions*. We will give a careful statement of a form of this theorem. Note that since we are working locally we operate in a normed linear space

E rather than an arbitrary manifold. The theorem requires E to be a Banach space, since the proof involves techniques of infinite iteration and needs good criteria for convergence of sequences.

THEOREM (Local existence and uniqueness of solutions of ordinary differential equations)

Let U *be an open set in a Banach space* E, *and let* $f : U \to E$ *be a* C^r *map with* $r \geq 1$. *Then given any point* p *in* U *there exists*

a positive number ε,

and *a neighbourhood* W *of* p *in* U

such that

for any x *in* W *and any positive number* $\delta \leq \varepsilon$

there exists a unique solution curve

$$c_x : (-\delta, \delta) \to U$$

to the equation $\dot{x} = f(x)$, *satisfying* $c_x(0) = x$.

Moreover, if we define a map

$$\phi : W \times (-\delta, \delta) \to U$$

by

$$\phi(x,t) = c_x(t)$$

then ϕ *is a* C^r *map. Thus solutions vary* C^r *with respect to* t *and the initial conditions.*

Remarks

1. This may seem rather an elaborate formulation of a simple idea. However, the reason for introducing W is that we want solutions to exist for $-\varepsilon < t < \varepsilon$ throughout a whole neighbourhood of p. We may not be able to use U itself for this neighbourhood, since the nearer x is to the edge of U the shorter the positive (say) t-interval over which c_x is defined

may be. See Figure 36. Also, we must formally allow the flexibility with

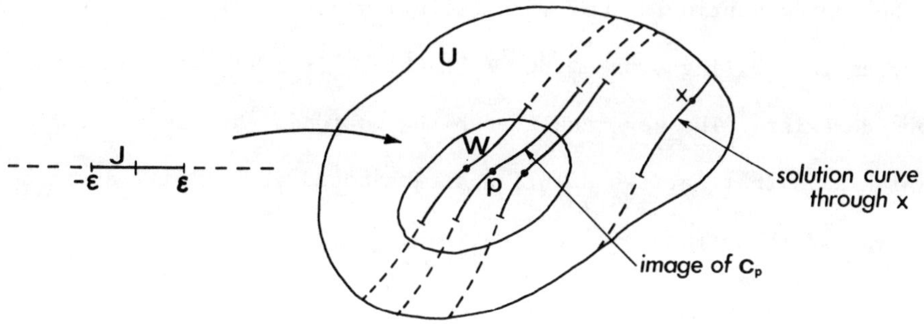

Figure 36

δ, since otherwise it is just conceivable that solutions c_x^1, c_x^2 could both exist with $c_x^i(0) = x$ $(i = 1,2)$ and be defined for, say, $-\varepsilon/2 < t < \varepsilon/2$ without contradicting the uniqueness statement for ε. (Of course these solutions would have to be non-extendible to solutions on $-\varepsilon < t < \varepsilon$, or ε-uniqueness *would* be contradicted.)

2. For proofs of the theorem see Dieudonné [32], Lang [67], or any of the many books on differential equations such as Arnol'd [11], Coddington and Levinson [30], Hartman [51], Lefschetz [69]. There is a neat proof using the Inverse Function Theorem directly, due to Robbin [104]. Most proofs begin by supposing X only to be C^1 or satisfying an even weaker *Lipschitz condition* (R. Lipschitz, 1832-1903), and use induction to prove the theorem in the C^r case. Moreover, proofs are usually given for the non-autonomous case, where f is a function of both x and t and the smoothness or Lipschitz conditions apply to x but not necessarily to t. However, we shall always be assuming smoothness (usually C^∞) in both variables and accordingly reduce the problem to the autonomous case as described on page 183.

We now want to extend this local existence and uniqueness theorem to a global theorem by continuing our local solution curves all over the manifold M as far as they will go, assuming X to be $C^r (r \geq 1)$ so that local solutions do exist. The key result enabling us to do this is the following lemma which says that any two solution curves based at p will agree on any interval on which they are both defined.

LEMMA

Let $c_1 : J_1 \to M$ *and* $c_2 : J_2 \to M$ *be two solution curves for* X *both based at the point* p *on* M. *Then* $c_1(t) = c_2(t)$ *for all* t *where* c_1 *and* c_2 *are both defined, i.e. on the intersection* $J_1 \cap J_2$.

Proof

This may seem obvious, but it is instructive to see how several topological notions from Chapter 1 are involved in formalizing this intuition. Let I be the subset of $J_1 \cap J_2$ on which c_1 and c_2 agree; we aim to prove $I = J_1 \cap J_2$. Note the following facts:-

(1) I is *not empty* (it contains 0);

(2) I is *open* in $J_1 \cap J_2$ (the local uniqueness guarantees this);

(3) I is *closed* in $J_1 \cap J_2$. (If t is not in I then $c_1(t) \neq c_2(t)$ and since M is Hausdorff we can find neighbourhoods of $c_1(t)$ and $c_2(t)$ which do not meet. The continuity of c_1, c_2 thus implies that $c_1(s) \neq c_2(s)$ for all s within some distance γ of t, i.e. the complement of I is open so I is closed.)

Now $J_1 \cap J_2$ is an interval, and is therefore connected (see §1.7). Hence (1), (2) and (3) imply that $I = J_1 \cap J_2$, proving the lemma.

Finally, we construct a maximal solution curve $c : J \to M$ based at p (i.e. one which cannot be extended over any larger t-interval containing J) as follows. Let $\{c_\lambda : J_\lambda \to M\}$ be the family of *all* solution curves based at p. Let J be the union of all the various intervals J_λ, and define $c : J \to M$ to be equal to c_λ on each J_λ. Then J is an interval (finite or infinite) and the above lemma shows that the definition of c is unambiguous. Clearly $c : J \to M$ is the unique maximal solution curve for X based at p. We call it the *global solution curve* through p.

Global solutions need *not* be defined for all time t:

EXAMPLES

1. The equation $\dot{x} = 1$ has global solution through p given by $c(t) = p+t$. If the equation is defined on the open interval $M = (-1,1)$ in $\mathbb{R}$, then each solution curve is defined in M only for a finite time.
2. Let $M = \mathbb{R}$ and consider the equation $\dot{x} = x^2$. The global solution through p is given by $c(t) = (p^{-1}-t)^{-1}$, but when $p > 0$ this solution is defined only for $-\infty < t < p^{-1}$.

The problems are that (1) $c(t)$ may reach the 'edge' of M in a finite time, or (2) $c(t)$ can shoot off to infinity while t approaches some finite number. However, if M is *compact* there is no edge (our manifolds here have no boundary) and no way of shooting off to infinity, and so we would expect global solutions to be defined in this case for all t. The optimism is justified.

THEOREM

If M is compact then each global solution curve for X is defined for all t in $\mathbb{R}$.

Proof

Let $c : J \to M$ be the global solution curve through p, and suppose $J = (a,b)$. We shall show $b = \infty$; the argument for $a = -\infty$ is similar. If $b < \infty$ we could find $t_1, t_2, \ldots$ with $t_n \to b$ as $n \to \infty$ (e.g. $t_n = b - \frac{1}{n}$). Look at the points $c(t_1)$, $c(t_2), \ldots$ in M.

If $c(t_m) = c(t_n)$ for some m,n, then local uniqueness and the above lemma show that c is periodic with $c(t) = c(t+t_n-t_m)$ valid for all t in $\mathbb{R}$, contradicting $b < \infty$. Hence all the $c(t_n)$ are distinct. By compactness of M there exists an accumulation point x_* of the $c(t_n)$'s (see §1.6). Now the local existence and uniqueness theorem shows that unique solutions exist based at any x in some neighbourhood W of x_* and defined for $-\varepsilon < t < \varepsilon$ where ε depends on W but not on x. Now choose t_n such that $b - t_n$ is less than ε and $c(t_n)$ is in W (possible by continuity of c). By local uniqueness and the lemma above we then have c extendible to a solution on $(a, t_n+\varepsilon)$, and since $t_n + \varepsilon > b$ we have contradicted the maximality of J. Therefore $b = \infty$, as originally claimed.

Thus if M is compact (and often when it is not) we can define $\phi(p,t) = c_p(t)$ where c_p is the global solution for X through p, and thereby obtain a map

$$\phi : M \times \mathbb{R} \to M.$$

This map, representing all the solutions for X simultaneously, has a number of interesting properties which will be of central importance to what follows. First of all, ϕ can be shown to be C^r if X is C^r, by piecing together the local information from the local existence and uniqueness theorem. Secondly, if we write $\phi_t(p) = c_p(t)$ then $\phi_t : M \to M$

satisfies

$$\phi_t \cdot \phi_s = \phi_{t+s} : M \to M$$

for all s,t in $\mathbb{R}$. This essentially says that uniqueness of solutions gives $t \mapsto c_p(t+s)$ as the unique global solution based at $c_p(s)$, so $\phi_{t+s}(p) = \phi_t(\phi_s(p))$. Also

$$\phi_0 = id_M$$

since $\phi_0(p) = c_p(0) = p$. It immediately follows that ϕ_t is a *diffeomorphism* because $\phi_t \cdot \phi_{-t} = \phi_{-t} \cdot \phi_t = \phi_0 = id_M$ so ϕ_{-t} is the inverse of $\phi_t : M \to M$.

Such a map ϕ is called a *flow* on M. This is the mathematical object which we shall use to model the general concept of a *dynamical system* governed by ordinary differential equations that we have been pursuing throughout the first three chapters of this book. The qualitative theory of dynamical systems becomes the theory of the *structure of flows on manifolds*, and this will occupy us for the whole of the fourth chapter.

Remarks on the literature. Over the last decade or so there has been a slow but steady stream of books becoming available on differential topology. In rough chronological order (relevant since language and aims of exposition tend to change) there are Lang [66], Munkres [86], Auslander and Mackenzie [13], Bishop and Crittenden [15], Spivak [128], Milnor [83], Wallace [145], Narasimhan [87], Spivak [129], Brickell and Clark [20], Lang [68], Golubitsky and Guillemin [42], Guillemin and Pollack [48], Dodson and Poston [33], Hirsch [52]. See also survey articles by Chillingworth [25], Stamm [130]. The formal theory of differentiable manifolds is recorded in Bourbaki [18]. The series of volumes [129] by Spivak provide a very thorough and lively account of manifolds and differential geometry, although unfortunately they are not always easy to obtain. For Morse theory see Milnor [82] or Wallace [145]; the excellent article by Palais [93] is also strongly recommended. Lie groups feature in some books on differential topology: see Bishop and Crittenden, Brickell and Clark, or Spivak [129]. The classic text is Chevalley [24]; other treatments are Hochschild [56] or the concise and very clear lecture notes by Adams [4]. A standard reference for theorems on the orbit structures of Lie group actions is Borel [17].

Some references for algebraic topology are Agoston [5], Hocking and Young [57], Maunder [79].

For the basic theory of differential equations refer to any of the texts quoted on page 185, and also the short introduction to some qualitative aspects of the theory by Hurewicz [59]. The books by Arnol'd [11] and Hirsch and Smale [55] are both written very much from the qualitative point of view, and contain large numbers of interesting examples, applications and fresh ideas: they would provide a stimulating accompaniment to these notes.

4 Qualitative theory of dynamical systems

4.1 FLOWS AND DIFFEOMORPHISMS

We saw in the previous chapter that the solutions of a C^r vector field (differential equation) on a compact manifold M give rise to a C^r *flow*

$$\phi : M \times R \to M$$

on M. Writing $\phi(p,t) = \phi_t(p)$, each ϕ_t is a diffeomorphism $M \to M$ and we have the relations

$$\left.\begin{aligned} \phi_0 &= id_M \\ \phi_t \cdot \phi_s &= \phi_{t+s} \end{aligned}\right\} \quad (*)$$

Let $\mathrm{Diff}^r(M)$ denote the set of all C^r diffeomorphisms $M \to M$. Then the composition of two elements of $\mathrm{Diff}^r(M)$ is another one, and under this operation of composition $\mathrm{Diff}^r(M)$ has the algebraic properties of a *group*. The relations (*) say that the map

$$R \to \mathrm{Diff}^r(M) : t \mapsto \phi_t$$

is a *group homomorphism*, or in the language of Lie groups (see §3.2 [12]) that the flow ϕ is an *action* of R on the manifold M. The images of the solution curves of the vector field are the *orbits* of this action.

If we fix a value t and consider positive and negative iterations

$$\ldots \phi_{-2t}, \phi_{-t}, \phi_0, \phi_t, \phi_{2t} \ldots$$

(remember $\phi_{-t} = (\phi_t)^{-1}$ and $\phi_{nt} = (\phi_t)^n$) we have a map

$$Z \to \mathrm{Diff}^r(M) : n \mapsto \phi_{nt}$$

which is an action of the group Z on M. This action is a kind of discrete approximation to the R-action ϕ, and we would expect it to give a more and more accurate picture of ϕ as we took smaller and smaller choices of t. On the other hand, given *any* diffeomorphism $f : M \to M$ we automatically have a Z-action on M by taking n to f^n, meaning f iterated n times, for each n in Z. We call this the Z-action *generated* by f. The set of points $\{f^n(x)\}$ for all $n = 0, \pm 1, \pm 2, \ldots$ is the *orbit* of the point x. It does not follow that f need come from any flow ϕ : see below.

In both the flow and diffeomorphism case the total configuration of all the orbits is called the *phase portrait*. It is this which we study when trying to understand the global qualitative behaviour of a dynamical system.

Remarks

1. One approach to the problem of classifying phase portraits is to take the quotient space (§1.3) of M by the equivalence relation R of 'lying in the same orbit', so that orbits are then represented by points in the *orbit space* M/R. However, this leads in general to such complications that it hardly sheds any light on the problem.
One difficulty is that M/R may easily be non-Hausdorff. Try to imagine, for example, the orbit space for an irrational flow on the torus (Example 6 below), or even for an innocent linear system in R^2 such as $\dot{x}_1 = x_1, \dot{x}_2 = -x_2$: see Figure 44.
2. In the same way that a flow represents the evolution of a system governed by ordinary differential equations, a diffeomorphism represents a *difference equation* of the form $x_{m+1} = f(x_m)$ where each x_m is a point on M. The sequence of points $x_0, x_1, x_2, \ldots$ is the forward orbit

of x_0 under f, and by including $x_{-1} = f^{-1}(x_0)$ and so on we obtain the whole orbit. Therefore the theory of diffeomorphisms as Z-actions is the global qualitative theory of certain types of difference equations. For many types of difference equation the map f may not be invertible, however, and so we no longer have a Z-action. Nevertheless, we could still study the behaviour of an arbitrary smooth map $f : M \to M$ under forward iterations. Although there are some interesting results along these lines, the global theory remains relatively unexplored.

3. We could drop differentiability from the discussions, and consider *continuous* actions where each ϕ_t or f is merely a homeomorphism: then we need not work with manifolds, but any topological spaces we liked. The theory of such actions (often called *continuous flows* when ϕ varies continuously with t, or *discrete flows*, respectively) is called *topological dynamics*. Many useful results about flows can be proved in this generality, but differentiability yields much stronger results. We shall rely heavily on differentiability, which can usually be assumed in practical applications.

The relationships between R*-actions and* Z*-actions*

There are two important ways in which R-actions give rise to Z-actions.

1. A choice of t gives the Z-action $n \mapsto \phi_{nt}$ from the R-action ϕ, as described above.

2. Let ϕ be a flow on M with corresponding vector field X, and suppose that Σ is a codimension-1 submanifold of M with the following properties:

(a) every orbit meets Σ for arbitrarily large positive and negative times t ;

(b) if p lies on $\sum$ then $X(p)$ is not tangent to $\sum$, i.e. $X(p)$ does not lie in $T_p\sum$.

Thus orbits intersect $\sum$ transversally.

Then $\sum$ is called a *cross-section* of the flow ϕ. It is not hard to prove that ϕ induces a Z-action on $\sum$, generated by $f : \sum \to \sum$ when for p on $\sum$ we define $f(p)$ to be $\phi_{t_1}(p)$ where t_1 is the least positive t for which $\phi_t(p)$ again lies in $\sum$. In general t_1 depends on p. We call f the *first-return map* or *Poincaré map* for $\sum$. See Figure 37. The flow on $M \times S^1$ arising from a periodic non-autonomous system on M (page 183) has a cross-section $\sum = M \times \{\theta\}$ for any fixed $\theta \in S^1$.

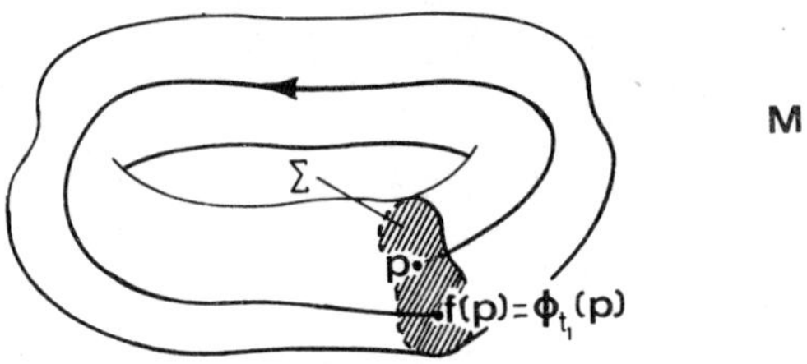

Figure 37

Conversely, a Z-action gives rise to an R-action in the following way. Let the Z-action be generated by f in $\text{Diff}^r(M)$. Take the product $M \times I$ (where I is the closed interval $[0,1]$) and take a unit vector field X pointing in the I-direction, thinking of $M \times I$ as contained in $M \times R$. Now glue the two ends of $M \times I$ together, attaching the point $(x,0)$ at the 0-end to the point $(f(x),1)$ at the 1-end. All this can be made respectable, using the idea of quotient topology (§1.3): compare the mathematical description of the Möbius band in §3.2, [9]. See Figure 38.

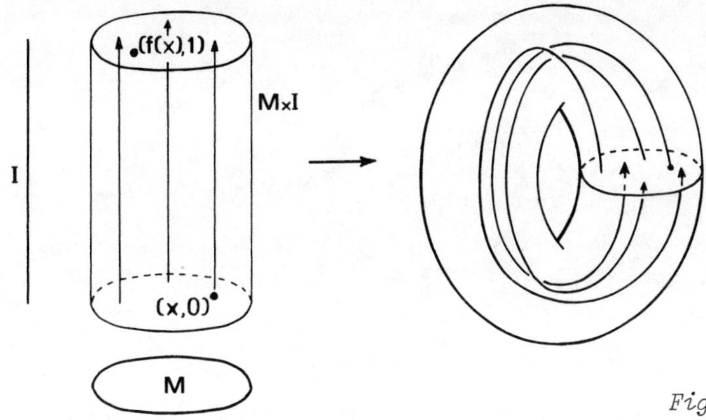

Figure 38

We obtain a vector field on a smooth manifold of dimension $n + 1$ (if $\dim M = n$), and the corresponding flow, called the *suspension* of f, has M as a cross-section.

Thus R- and Z-actions can be associated to each other in ways which reflect their orbit-structures. This association has serious limitations, though:

1. *Not every flow is the suspension of a diffeomorphism,* since for example a suspension has no points where the vector field vanishes. This also implies that not every flow has a cross-section.
2. *Not every diffeomorphism can be incorporated into a flow on the same manifold* (i.e. as ϕ_t for some t). Clearly a diffeomorphism which is not deformable continuously to the identity cannot be ϕ_t for any t. Deformability is still not enough: let $f : S^1 \to S^1$ be a diffeomorphism defined by $f(e^{i\theta}) = \rho \cdot \psi_1(e^{i\theta})$, where ψ is a flow whose orbits are as illustrated in Figure 39, and ρ is rotation by 180°.

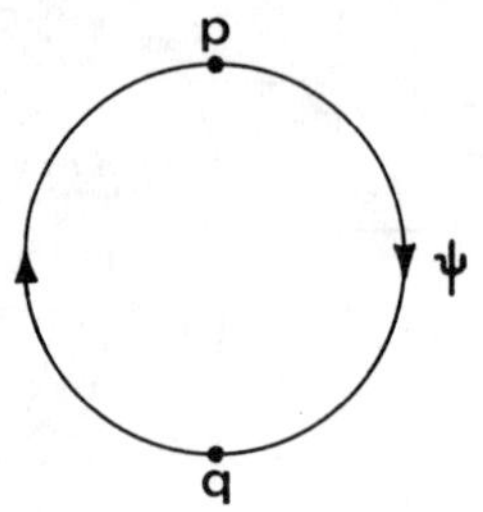

Figure 39

Then $f(p) = q$, $f(q) = p$, so $f^2(p) = p$, $f^2(q) = q$. If $f = \phi_t$ for some flow ϕ, then the whole circle must be one ϕ-orbit (or p,q would not be periodic) which would have to be periodic of period $2t$: see below. But this is false since $f^2(r) \neq r$. Therefore f is not of the form ϕ_t for any ϕ and t.

Nevertheless, it is clear that the study of flows and the study of iterated diffeomorphisms are very closely related, and we shall pursue them both in parallel. For flows we tend to write $\phi_t(p)$ as $\phi_t p$.

Important features

1. A *fixed point* of a flow ϕ is a point p for which $\phi_t p = p$ for all t, i.e. for which the vector field vanishes: $X(p) = 0$. Thus it is also called a *zero* (or, misleadingly, a *singularity* or *critical point*) of the vector field. It corresponds to an *equilibrium* state of the dynamical system being modelled.

A *fixed point* of a diffeomorphism f is a point p for which $f(p) = p$, and hence $f^n(p) = p$ for all integers n.

2. A *periodic point* of a flow ϕ is a point q for which there exists some $T > 0$ with $\phi_T q = q$. The least such T is called the *period* of q. If $r = \phi_s q$ is any point on the orbit of q then

$\phi_T r = \phi_T \phi_s q = \phi_{T+s} q = \phi_{s+T} q = \phi_s \phi_T q = \phi_s q = r$, so r is also a periodic point with the same period as q. Therefore we describe the orbit γ of q as a *periodic orbit* of period T.

A *periodic point* of a diffeomorphism f is a point q such that $f^m(q) = q$ for some $m > 0$ (or > 1 if we want to distinguish periodic points from fixed points). The least such m is the *period* of q.

Every fixed or periodic point of f gives rise to a periodic orbit of the suspension flow; conversely, every periodic orbit of a flow with a cross-section $\sum$ gives rise to a fixed or periodic point of the first-return map for $\sum$. This fact that the phenomenon of periodicity is preserved when taking suspensions and cross-sections is one of the justifications for studying diffeomorphisms in order to understand flows.

Fixed points and periodic orbits correspond to 'observable' phenomena in a dynamical system, since they represent states which are unchanging or which repeat themselves. However, points whose orbits return infinitely often to any arbitrarily small neighbourhood of them, or even (since we can only measure to within a certain finite accuracy) points p which have points q arbitrarily close to p which return infinitely often (for times $t \to \infty$) to any neighbourhood of p, may also be associated to some kind of observable phenomenon although it could have a rather chaotic or turbulent character. Recurrent behaviour of this type is called non-wandering behaviour. A point p is *wandering* if it has a neighbourhood which never returns to intersect itself, and a point which is not wandering is *non-wandering*. Formally, we express this as follows:

DEFINITION

A point p *is a* non-wandering point *for the flow* ϕ *on* M *if, given any neighbourhood* U *of* p, *there exist arbitrarily large* $t > 0$ *for which* $U \cap \phi_t V$ *is non-empty. Similarly for a diffeomorphism* f, *with* f^n $(n > 0)$ *replacing* ϕ_t .

Remarks

1. The set of non-wandering points is denoted by $\Omega(\phi)$, $\Omega(f)$ or simply Ω .

2. Let $V = \phi_s U$. Then if $\phi_t U \cap U$ is nonempty it follows that $\phi_s(\phi_t U \cap U) = \phi_{s+t} U \cap \phi_s U = \phi_t V \cap V$ is also nonempty. From this it is easy to deduce

(i) if x is in $\Omega(\phi)$ then the whole orbit of x is in $\Omega(\phi)$;

(ii) using $t < 0$ would give the same definition of $\Omega(\phi)$ (put $s = -t$ in the above).

Result (ii) shows that if $\bar{\phi}$ is the flow obtained from $-X$ then $\Omega(\bar{\phi}) = \Omega(\phi)$.

Similar remarks apply to $\Omega(f)$. In particular, $\Omega(f) = \Omega(f^{-1})$.

3. Clearly fixed points and periodic orbits all lie in Ω.

4. The set of wandering points is obviously an open set, and so Ω is a closed set. If M happens to be compact this will imply that Ω is compact. The fact that Ω is closed means that it must contain at least the *closure* of the set of fixed and periodic points.

5. If a point is wandering it corresponds to transient behaviour in a dynamical system, and so to understand the long-term behaviour we need to know where (if anywhere) the point can be said to approach asymptotically

as $t \to +\infty$. For any point x we say that another point y is an *ω-limit point* of (the orbit of) x if there are points $\phi_{t_1}x, \phi_{t_2}x, \ldots$ on the orbit of x with $t_i \to +\infty$ and $\phi_{t_i}x \to y$ as $i \to \infty$. This does *not* necessarily imply that $\phi_t \to y$ as $t \to +\infty$: see examples below. The set of all such y for given x is called the *ω-limit set* of x, written $\omega(x)$.

Replacing $t \to +\infty$ by $t \to -\infty$ gives the *α-limit set* $\alpha(x)$.

Replacing t_i by n_i gives a corresponding definition of α and ω for diffeomorphisms.

A straightforward argument shows that $\omega(x)$ and $\alpha(x)$ are both subsets of Ω, for every x. If x is fixed or periodic then $\alpha(x) = \omega(x) =$ orbit of x.

EXAMPLES of flows

1. *Simple harmonic oscillator* $\ddot{x} + k^2x = 0$

We convert this second order equation on R into a first order system on $TR = R \times R = R^2$ by setting $\dot{x} = y$ and obtaining

$$\dot{x} = y, \quad \dot{y} = -k^2x \quad .$$

The equation is easy to solve explicitly, and a formula for the flow $\phi : R^2 \times R \to R^2$ can be worked out as

$$\phi(x,y;t) = (x \cos kt + k^{-1}y \sin kt, \; -kx \sin kt + y \cos kt) \quad .$$

(It is instructive to verify that this *is* a flow.) The orbits are ellipses $k^2x^2 + y^2 =$ constant, except for $(0,0)$ which is the unique fixed point. Every non-fixed point is periodic, so every point is non-wandering: $\Omega = R^2$.

2. *Damped simple harmonic oscillator* $\ddot{x} + 2b\dot{x} + k^2x = 0$ $(b > 0)$

Here $\dot{x} = y$, $\dot{y} = -2by - k^2x$ and *without working out the equations for the flow* we know that the vectors now have an 'inwards' component and so we deduce on qualitative grounds alone that all orbits spiral in towards the origin as $t \to +\infty$. Hence the only ω-limit set is the origin. Also, every $\alpha(p)$ is empty except that $\alpha(0,0) = (0,0)$. Thus Ω here consists of the origin and nothing else.

This and the previous example are both *linear* systems, meaning that the map $R^2 \to R^2 : (x,y) \mapsto (\dot{x},\dot{y})$ is a linear map. For pictures of the orbit structures of these and other linear systems in R^2 see Figure 44.

3. *Simple pendulum.* $m\ell\ddot{\theta} + mg \sin\theta = 0$

Take $\ell = g = 1$, so θ represents a point on S^1, and convert this second order equation on S^1 into a first order equation on $TS^1 = S^1 \times R$ by putting $\dot{\theta} = v$ to obtain $\dot{\theta} = v$, $\dot{v} = -\sin\theta$. When $S^1 \times R$ is 'unrolled' the orbits of the flow are as shown in Figure 40.

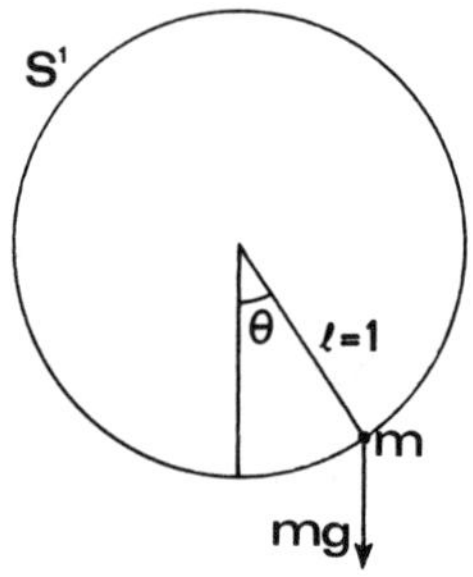

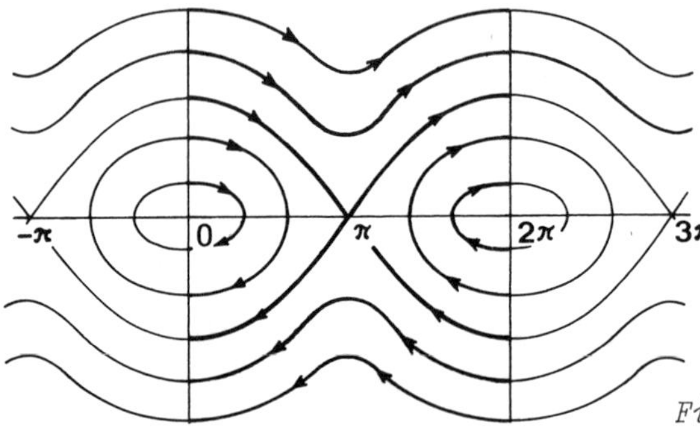

Figure 40

Each orbit lies in some subset $E^{-1}(c)$ where $E : TS^1 \to R$ is the *energy* map $(\theta,v) \mapsto \frac{1}{2}v^2 - \cos\theta$ (this simply says that energy is conserved): see Figure 41. Here Ω is all of TS^1.

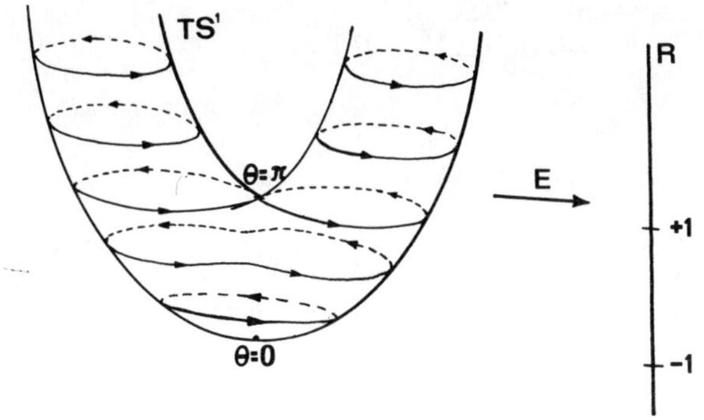

Figure 41

Observe that E has two critical points (0,0) and $(\pi,0)$; they are both non-degenerate (minimum and saddle-point, respectively). For every regular value c of E the inverse image $E^{-1}(c)$ is a codimension-1 submanifold of TS^1 (see §3.3, page 157). It is now a very instructive exercise to visualize what happens to the orbits when a damping term is added, turning the equation into

$$\ddot{\theta} + b\dot{\theta} + \sin\theta = 0$$

and corresponding to a loss of energy along each non-fixed orbit.

4. *Gradient flows on* R^n

If X is a vector field on R^n corresponding to equations of the form

$$\dot{x}_1 = \frac{\partial V}{\partial x_1}, \; \dot{x}_2 = \frac{\partial V}{\partial x_2}, \; \ldots, \; \dot{x}_n = \frac{\partial V}{\partial x_n}$$

where $V : R^n \to R$ is some smooth function, then X is called a *gradient* vector field. It is easy to show that V can never decrease along solution curves, since

$$\frac{d}{dt} V(c(t)) = DV(c(t)).\dot{c}(t)$$

$$= \sum_{i=1}^{n} \frac{\partial V}{\partial x_i} (c(t)).\dot{c}_i(t)$$

$$= \sum_{i=1}^{n} (\dot{c}_i(t))^2 \geqslant 0 \quad .$$

If fact $\frac{d}{dt} V(c(t)) > 0$ unless $\dot{c}(t) = 0$, in which case (by uniqueness of solutions) $c(t) =$ constant. From this it follows that if solutions are defined for all t, so that X gives rise to a flow ϕ on R^n, *the only non-wandering points are fixed points* - i.e. there can be no periodic or other recurrence phenomena: this is because once a point $c(t) = \phi_t(p)$ has left $p = c(0)$ the value of V must have increased and so by the continuity of V there will be some neighbourhood of p into which $\phi_t(p)$ can never return.

This relatively simple structure of Ω for gradient flows on R^n allows their qualitative behaviour to be analyzed quite thoroughly. We shall return to this important fact and its implications later, in the wider context of manifolds.

5. *The van der Pol oscillator* $\ddot{x} - \alpha(1-x^2)\dot{x} + x = 0$, $\alpha > 0$

Converting this as usual into a first order system, we have

$$\dot{x} = y$$
$$\dot{y} = \alpha(1-x^2)y - x$$

which is a non-linear system exhibiting a particularly important type of behaviour. The origin is clearly a fixed point, and local analysis there (see §4.2) shows that solution curves are spiralling away from the origin. However, they do not spiral out to infinity but approach a periodic orbit γ which surrounds the origin. Outside γ the solution curves spiral in towards γ, and γ is therefore called a *limit cycle*. It is the ω-limit set of every point except the origin; here $\Omega = \gamma \cup \{\text{origin}\}$. See Figure 42.

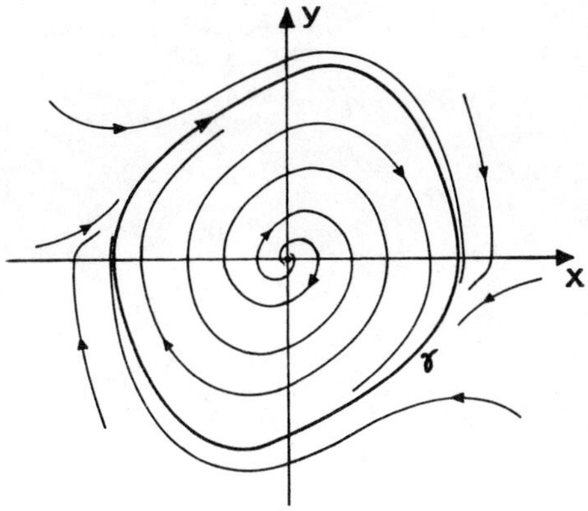

Figure 42

The physical interpretation of the picture is that, no matter what the initial conditions are apart from (0,0), the system will eventually (or possibly very quickly) be behaving in a way which is practically indistinguishable from the periodic behaviour of points on γ. Indeed, even if it is initially at (0,0), the slightest perturbation can cause it to rocket out towards γ.

It is surprisingly difficult to prove rigorously the existence of the single limit cycle γ for this system, although it is easy enough to construct artifically other examples of systems exhibiting limit cycles. For example

$$\begin{aligned} \dot{x} &= y + (1-r^2)x \\ \dot{y} &= -x + (1-r^2)y \end{aligned} \qquad (r^2 = x^2+y^2)$$

has the circle $r = 1$ as a limit cycle. The van der Pol equation is historically important as the basic equation for the oscillation of a radio valve.

6. *Rational and irrational flows on the torus*

Recall that in §3.3 we studied immersions $R \to S^1 \times S^1$ of the form $t \mapsto (\exp i\alpha t, \exp i\beta t)$. Such a curve arises as a solution curve to the vector field on $S^1 \times S^1$ corresponding in (θ,ϕ) 'co-ordinates' to the first order system

$$\dot{\theta} = \alpha$$

$$\dot{\phi} = \beta$$

(in fact it is the global solution through $(\theta,\phi) = (0,0)$). If α/β is rational then *every* orbit of the corresponding flow is periodic, hence non-wandering. If α/β is irrational, then every orbit is dense in the torus, hence again non-wandering. Therefore in both cases $\Omega = S^1 \times S^1$.

4.2 LOCAL BEHAVIOUR NEAR FIXED POINTS AND PERIODIC ORBITS

In studying the qualitative properties of flows and diffeomorphisms our approach will be (1) to investigate behaviour on and near the non-wandering set Ω (where the action is) and then (2) piece together a global picture from information about how orbits go from (near) one part of Ω to (near) another. We begin by looking at behaviour near fixed points, corresponding physically to behaviour near equilibrium states. First we will consider flows, and then discuss the analogous behaviour for diffeomorphisms.

Local linearization of flows

Let p be a fixed point of the smooth flow ϕ on M. Since $\phi_t p = p$ we have a linear map

$$T_p\phi_t : T_pM \to T_pM$$

for every t, and the Chain Rule for tangent maps (page 166) gives

$$T_p\phi_t \cdot T_p\phi_s = T_p(\phi_t \cdot \phi_s) = T_p\phi_{t+s}$$

so as t runs through R the family $\{T_p\phi_t\}$ of (linear) diffeomorphisms gives us a flow on the linear space T_pM. We call this the *linearization* of ϕ at p.

If ϕ arises from the vector field X on M then by taking a co-ordinate chart on a neighbourhood U around p we can represent X as a system

$$\dot{x} = X(x)$$

in R^n, where (see the beginning of §3.5) the symbol is now used to denote a map $X : U \to R^n$. By translating the origin we can suppose that p corresponds to 0 in R^n, so $X(0) = 0$. The derivative $DX(0)$ of X at 0 is a linear map $R^n \to R^n$, and we can consider the linear system

$$\dot{x} = DX(0).x$$

on R^n. This is called the *linearization* of X at 0.
Not surprizingly, the flow corresponding to this linear system turns out to be the coordinate representation in U of the above linearization of ϕ at p. The proof is simple: if X gives rise to the flow ϕ in R^n then $\frac{\partial}{\partial t}\phi_t x = X(\phi_t x)$ and so differentiation with respect to x gives

$$\frac{\partial}{\partial t} D\phi_t(x) = DX(\phi_t x) \cdot D\phi_t(x)$$

since we can interchange differentiation with respect to x and t (see page 77). Now we put $x = 0$ and get

$$\frac{\partial}{\partial t} D\phi_t(0) = DX(0) \cdot D\phi_t(0)$$

showing that $(v,t) \mapsto D\phi_t(0)v$ is the flow arising from $DX(0)$.

It is reasonable to ask whether knowing about the behaviour of the linearization of ϕ at p will be of any use in trying to understand the behaviour of ϕ itself near p. As with the Inverse Function Theorem and the Morse Lemma for maps and for functions, the answer is Yes, provided certain non-degeneracy conditions are met.

Since we are working locally we may as well suppose that $M = \mathbb{R}^n$ and $p = 0$, and we will once again use the language of *germs* (§2.7). The vector field X is thought of locally as the germ at 0 of a smooth map $X : \mathbb{R}^n \to \mathbb{R}^n$. The kind of equivalence we are interested in this time will not be right equivalence or right-left equivalence but something specially adapted to vector fields, since a local change of co-ordinates in the domain of X automatically gives a transformation of both the right hand side and the left hand side of the system $\dot{x} = X(x)$. Explicitly, if $x = h(y)$ where h is a local diffeomorphism then $\dot{x} = Dh(y)\dot{y}$ and so the system in the new y-coordinates becomes

$$\begin{aligned}\dot{y} &= (Dh(y))^{-1}\dot{x} = Dh^{-1}(h(y)).X(h(y)) \\ &= Y(y), \text{ say.}\end{aligned}$$

Therefore the idea of two vector field germs being *the same up to differentiable* (C^r) *change of co-ordinates* becomes formally the following:

DEFINITION

Two vector field germs X, Y *are* C^r-equivalent *if there exists a* C^r *diffeomorphism germ* h *such that*

$$Y(\cdot) = Dh^{-1}(h(\cdot)).X(h(\cdot))$$

or, more simply,

$$Yh^{-1}(\cdot) = Dh^{-1}(\cdot).X(\cdot) \quad .$$

Now we can ask whether the germ of a vector field X at 0 is C^r equivalent for some r to the germ at 0 of the linearized vector field DX(0), and the answer we receive is: usually.

THEOREM (Sternberg [131])
For each integer $r \geq 2$ *there is a finite number of relationships among the eigenvalues of* DX(0) *with the property that* X *is* C^r*-equivalent to* DX(0) unless *one or more of the relationships holds.*

Remarks

1. The relationships are as follows: an eigenvalue λ_j can be expressed in terms of all the eigenvalues $\lambda_1, \lambda_2, \ldots, \lambda_n$ (including λ_j) as

$$\lambda_j = m_1\lambda_1 + m_2\lambda_2 + \ldots + m_n\lambda_n$$

where the $m_i \geq 0$ are integers with $2 \leq \sum m_i \leq$ a certain N(r). If any λ_j is purely imaginary ($\lambda_j = i\omega$) then $-\lambda_j$ is also an eigenvalue $\lambda_{j'}$, and so we can write

$$\lambda_j = 2\lambda_j + \lambda_{j'}$$

which is a relationship of the proscribed type. In this case X may not be C^1 linearizable. As we see on the next page, there are elementary and intuitive reasons why this is so, but it is important to realize that these Sternberg conditions are rather subtle and in general non-intuitive, showing for example that if $n = 2$ and $\lambda_1 = 1$, $\lambda_2 = 2$ then X may not be C^1 linearizable.

2. The set of linear maps $L : R^n \to R^n$ avoiding the Sternberg relationships for given r is open and dense in $L(R^n, R^n)$.

3. For X to be C^∞ linearizable it must avoid the Sternberg relationships for *every* $r \geq 1$, so an infinite number of conditions have

to be met. The set of linear maps $L : R^n \to R^n$ avoiding the relationships for *all* r is dense but no longer open in $L(R^n,R^n)$: it is a residual set (see §4.4).

If we are concerned with a qualitative understanding of the local behaviour of a dynamical system near an equilibrium, then Sternberg's Theorem is too powerful. For example, if we want to know if the equilibrium point p is a stable equilibrium (the orbits of points near p all approach p as $t \to +\infty$) we do not need C^r information about X but only topological (C^0) information about the configuration of the orbits of the flow locally. The key result here is the theorem of Hartman [50] and Grobman [44], usually referred to in non-Soviet literature as Hartman's Theorem. It was proved at a surprisingly late date in the development of the theory of differential equations.

THEOREM (Hartman)
If DX(0) *has no purely imaginary (including zero) eigenvalues, then there is a homeomorphism germ at* 0 *in* R^n *locally taking orbits of* ϕ *to orbits of the linearized flow (i.e. the flow defined by* DX(0)*). The homeomorphism preserves the sense of orbits* $(t \to +\infty$ *or* $-\infty)$ *but not necessarily the parametrization of orbits by* t.

That the eigenvalue condition is necessary is illustrated by the simple harmonic motion system $\dot{x} = y$, $\dot{y} = -x$ (see §4.1, Example 1). Here the field is already linear, but if we take X given by

$$\dot{x} = y + x^2 + y^2$$
$$\dot{y} = -x + x^2 + y^2$$

then all orbits except (0,0) spiral outwards and so in no neighbourhood of the origin can there be a homeomorphism taking orbits of this flow to the circular orbits of the simple harmonic flow arising from $DX(0,0)$.

Analogous results for diffeomorphisms

If a C^r diffeomorphism germ h converts the system $\dot{x} = X(x)$ into $\dot{y} = Y(y)$ then since we are after all dealing only with a change of co-ordinates it is clear that under h the orbits of the respective flows ϕ, ψ must agree and also respect parametrization by t, so that $h\phi_t(x) = \psi_t h(x)$ locally or, as germs,

$$h\cdot\phi_t = \psi_t\cdot h .$$

This says that the diffeomorphisms ϕ_t, ψ_t are locally *conjugate* (page 26) by the C^r diffeomorphism h. This notion of conjugacy at the germ level is what corresponds in the diffeomorphism case to C^r-equivalence in the vector field or flow case.

Sternberg's Theorem for local C^r linearization of a diffeomorphism germ f at 0 is as before, with $Df(0)$ replacing $DX(0)$ and C^r-conjugacy replacing C^r-equivalence. The relationships which the eigenvalues $\mu_1,\mu_2,\ldots,\mu_n$ of $Df(0)$ should avoid are

$$\mu_j = \mu_1^{m_1}\,\mu_2^{m_2}\,\cdots\,\mu_n^{m_n}$$

for $m_i \geqslant 0$ and $2 \leqslant \sum m_i \leqslant N(r)$.

Hartman's Theorem becomes the following:

THEOREM (Hartman's Theorem for diffeomorphisms)

If $Df(0)$ *has no eigenvalues with modulus equal to* 1 *then the germ of* f *at* 0 *is* C^0*-conjugate to the germ at* 0 *of the linear map* $L = Df(0)$.

It follows immediately that the conjugating homeomorphism germ h satisfying $hf = Lh$ will locally take the orbit of each x under the f-action to the orbit of $h(x)$ under the L-action, since

$$h \cdot f^2(x) = h \cdot f \cdot f(x) = L \cdot h \cdot f(x) = L^2 \cdot h(x)$$

and similarly $h \cdot f^m(x) = L^m \cdot h(x)$ for every integer m.

Remarks

1. A typical *linear* system on $\mathbb{R}^n$ is of the form

$$\dot{x} = Ax$$

where A is a linear map $\mathbb{R}^n \to \mathbb{R}^n$, i.e. $A \in L(\mathbb{R}^n,\mathbb{R}^n)$. It is not hard to prove that the corresponding flow on $\mathbb{R}^n$ is given explicitly by the formula

$$\phi(x,t) = e^{tA}x$$

where e^{tA} is the element

$$I + tA + \frac{t^2}{2!}A^2 + \frac{t^3}{3!}A^3 + \ldots$$

of $L(\mathbb{R}^n,\mathbb{R}^n)$: the infinite series can be shown to converge.
Now the eigenvalues of e^{tA} are $\{e^{t\lambda_i}\}$ if $\{\lambda_i\}$ are the eigenvalues of A, and this provides the link between the two eigenvalue conditions for flows and diffeomorphisms respectively, namely that a linear vector field has a purely imaginary or zero eigenvalue precisely when the corresponding ϕ_t (for any $t \neq 0$) has an eigenvalue of modulus 1.
2. A linear isomorphism $L : \mathbb{R}^n \to \mathbb{R}^n$ having no eigenvalues of modulus 1 is called *hyperbolic*. By extension, a fixed point p of a diffeomorphism (germ) f is said to be *hyperbolic* if the linear map $T_pf : T_pM \to T_pM$ is hyperbolic: the verification of this condition does not depend on the choice of local co-ordinates. In view of Remark 1, a zero p of a

vector field X is called *hyperbolic* if the linearized vector field at p has no purely imaginary or zero eigenvalue; i.e. if each linearized flow map ϕ_t $(t \neq 0)$ is hyperbolic at the fixed point p.

3. There is an infinite-dimensional version of Hartman's Theorem which in its diffeomorphism form is a very valuable tool for proving stability theorems of a global type as in §4.5. Here the eigenvalue condition becomes a condition on the *spectrum* of the linear map. See Nitecki [90] for a proof, which is of course also applicable in finite dimensions.

To summarize: *near a hyperbolic fixed point a flow or diffeomorphism is topologically equivalent to its own linearization.* Having established this, we must now find out what hyperbolic linear flows and diffeomorphisms look like.

Structure of hyperbolic linear diffeomorphisms and flows

If $L : \mathbb{R}^n \to \mathbb{R}^n$ is a hyperbolic linear isomorphism then there are two linear subspaces E^+ (corresponding to eigenvalues λ with $|\lambda| > 1$) and E^- (corresponding to eigenvalues λ with $|\lambda| < 1$) such that $\mathbb{R}^n = E^+ \oplus E^-$ (see page 42) with the following properties:

each $E^{\pm}$ is invariant under L, i.e. $L(E^{\pm}) = E^{\pm}$;

all eigenvalues of $L \mid E^+$ have modulus > 1 ;

all eigenvalues of $L \mid E^-$ have modulus < 1 .

Furthermore, there exists a norm on $\mathbb{R}^n$ for which there exist constants $k < 1 < K$ such that

$$x \in E^+ \text{ implies } ||L(x)|| > K||x||$$

$$x \in E^- \text{ implies } ||L(x)|| < k||x|| ,$$

i.e. L expands along E^+ and contracts along E^- .

We can think of R^n as being decomposed into the direct sum of the two 'co-ordinate' subspaces: E^+ and E^-. The possibility that E^+ or $E^- = \{0\}$ is not excluded. If $E^+ = \{0\}$ then $E^- = R^n$ and we say the origin is a *sink*; if $E^+ = R^n$ then the origin is a *source*. Intermediate cases are often called *saddle points*. See Figure 43.

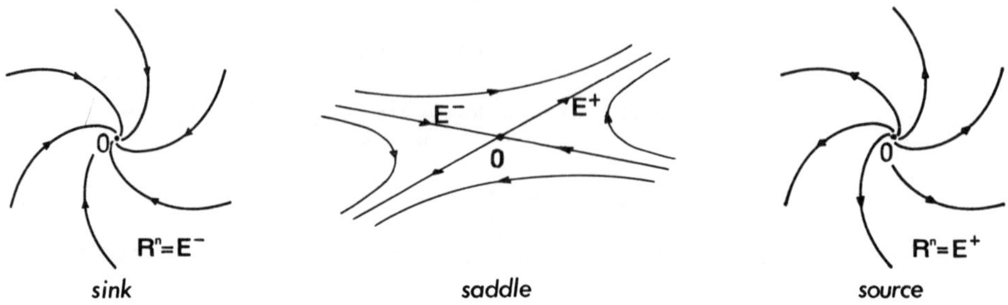

Figure 43

The proofs of the above assertions are straightforward matrix theory, although the construction of the norm is not entirely trivial. Analogous results apply in infinite dimensions, using the spectrum instead of the eigenvalues.

For a hyperbolic linear flow ϕ arising from $\dot{x} = Ax$ we have corresponding behaviour: R^n splits into $E^+ \oplus E^-$ such that each $E^\pm$ is invariant under the flow (i.e. $\phi_t E^\pm = E^\pm$) and all eigenvalues of $A \mid E^+$ $(A \mid E^-)$ have positive (negative) real part. There is a norm and constants $k < 1 < K$ such that

$$x \in E^+ \text{ implies } ||\phi_t x|| > K^t ||x||$$

$$x \in E^- \text{ implies } ||\phi_t x|| < k^t ||x||$$

for all $t > 0$. The situation can again be symbolized by Figure 43.

In two dimensions the orbits of hyperbolic linear flows can be sketched quite easily, since the corresponding system of equations can be solved explicitly: see Figure 44. Note that the validity of these pictures rests on the fact that if a hyperbolic linear system is perturbed linearly by a small amount so that no eigenvalue crosses the imaginary axis, then *the phase portrait remains the same up to homeomorphism.* This observation, although not mentioned again explicitly, is a first step in the theory of global stability discussed in §4.5 and should be kept well in mind.

Pictures of orbits of flows arising from 2-dimensional hyperbolic linear systems $\dot{x} = Ax$.

λ_1, λ_2 *real, negative*

(i) *equal, with two independent eigenvectors*

(ii) *unequal*

$\lambda_1, \lambda_2 = \alpha \pm i\beta$,

$\beta \neq 0, \quad \alpha < 0$

(iii) *or* (iv), *depending on sign of lower left hand element in* A

(v) $\lambda_1 = \lambda_2$ *real, negative with no two independent eigenvectors (A cannot be diagonalized)*

(vi) λ_1, λ_2 *real, opposite signs*

(i)-(v) *are* sinks, (vi) *is a* saddle-point. *For* sources *change signs of (real parts of)* λ_1, λ_2 *in* (i)-(v).

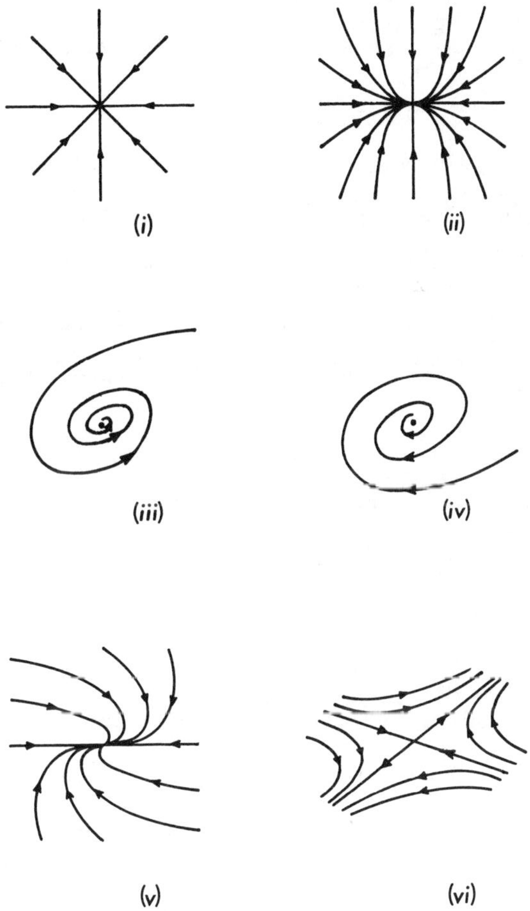

Figure 44

The discussion of hyperbolicity so far has been concerned with fixed points of flows and diffeomorphisms. It is only a small step to extend this to periodic orbits. If q is a periodic point with period m of the diffeomorphism f then we say that q is *hyperbolic* if it is hyperbolic as a fixed point of f^m. It then follows automatically that each point $f(q), f^2(q)\ldots$ on the orbit of q is also hyperbolic, so we can talk about a *hyperbolic periodic orbit* of f.

For a periodic orbit γ of a flow ϕ we need a different method of approach. Choose a point q on γ, and take a (small) piece of codimension-1 submanifold Σ passing through q and transverse to γ, i.e. such that the vector $X(q)$ does not lie in $T_q\Sigma$. For each x on Σ sufficiently close to q the continuity of the flow guarantees that the forward orbit of x will meet Σ again at a time t approximately equal to the period of γ. More precisely, there exists a neighbourhood V of q in Σ on which we can define a map

$$P : V \to \Sigma$$

by $P(x) = \phi_\tau x$ where τ (depending on x) is the smallest $t > 0$ for which $\phi_t x \in \Sigma$. Thus Σ is a kind of local cross-section, with P the first-return map (cf. page 194). Using the Implicit Function Theorem it can be shown that P is C^r if X is. Clearly $P(q) = q$, and the germ of P is a diffeomorphism germ at q, since we construct an inverse by following orbits backwards instead of forwards. Therefore we can discuss the hyperbolicity or otherwise of P as a diffeomorphism germ $\mathbb{R}^{n-1} \to \mathbb{R}^{n-1}$. If P is hyperbolic then we say that γ is a *hyperbolic periodic orbit* of the flow. This does not depend on the choice of Σ transverse to γ, nor on the choice of q on γ at which to construct Σ.

The germ at q of the map P is called a *Poincaré map* germ for γ. Recall that if $f : M \to M$ is any diffeomorphism then f^m is a (global) Poincaré map for every periodic orbit of the suspension of f (page 195) arising from a periodic point of f of period m.

Now we can apply Hartman's theorem to the germ of f^m in the diffeomorphism case or to the germ of P in the flow case, and it tells us that *the behaviour of a diffeomorphism or flow near a hyperbolic periodic orbit is topologically the same as the behaviour of the linearized version.*

In view of Hartman's Theorem and our analysis of hyperbolic linear systems above we now see that if we take a sufficiently small neighbourhood U of a hyperbolic fixed point p and define the local *in-set* of p in the diffeomorphism case by

$$\underline{\text{in}}(p) = \{x \in U \mid f^n(x) \to p \quad \text{as} \quad n \to +\infty\}$$

or in the flow case by

$$\underline{\text{in}}(p) = \{x \in U \mid \phi_t x \to p \quad \text{as} \quad t \to +\infty\}$$

then $\underline{\text{in}}(p)$ is homeomorphic to a neighbourhood of 0 in the linear subspace E^- of $\mathbb{R}^n$ defined from linearization of f or ϕ at p, so $\underline{\text{in}}(p)$ is a topological manifold of dimension s where s is the number of eigenvalues with modulus less than 1 (or real part < 0). In fact we can do better than this:

THEOREM

If the diffeomorphism or flow is C^r *then* $\underline{\text{in}}(p)$ *is a* C^r *submanifold of* M. *Its tangent space at* p corresponds to E^- in $\mathbb{R}^n$.

The terminology *in-set* was suggested by E.C. Zeeman. More traditionally $\underline{\text{in}}(p)$ is called a *local stable manifold* for p, denoted by $W^s_{\ell oc}(p)$. Here s means *stable*, and also refers to the dimension of

$W^s_{\ell oc}(p)$. Note that if we choose U carefully (corresponding to a small ball with respect to the relevant norm on R^n) then $fW^s_{\ell oc}(p) \subset W^s_{\ell oc}(p)$; similarly for ϕ_t with $t > 0$.

The proof of the theorem is not easy: see Smale [122] Nitecki [90], or Irwin [61]. It is known as the Stable Manifold Theorem.

Replacing n by $-n$ (or t by $-t$) we obtain the local *out-set* or *unstable manifold* $W^u_{\ell oc}(p)$ for p. It is a C^r submanifold of M of dimension $u = n - s$, with tangent space at p corresponding to E^+. For appropriate U we have $fW^u_{\ell oc}(p) \supset W^u_{\ell oc}(p)$; similarly for ϕ_t, $t > 0$.

The extension of these ideas to periodic orbits is straightforward. If q is a hyperbolic periodic point of period m of f then as above we have local stable and unstable manifolds for f^m at q. Transporting these around the orbit by applying $f, f^2, \ldots, f^{m-1}$ we obtain local stable and unstable manifolds for $f(q), f^2(q), \ldots, f^{m-1}(q)$: see Figure 45.

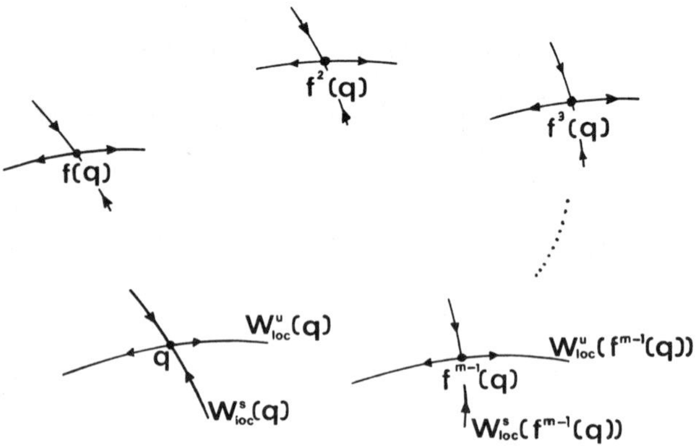

Figure 45

We can think of these m stable (unstable) manifolds as together constituting a stable (unstable) manifold for the periodic orbit as a whole.

Similarly, if γ is a hyperbolic periodic orbit for the flow ϕ then the local stable and unstable manifolds for a Poincaré map can be transported around γ by ϕ to give manifolds of one higher dimension as local stable and unstable manifolds $W^s_{\ell oc}(\gamma)$, $W^u_{\ell oc}(\gamma)$ for γ as a whole: see Figure 46.

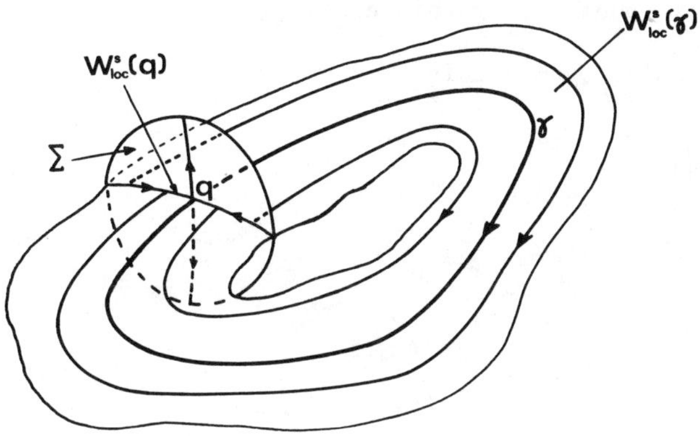

Figure 46

4.3 SOME GLOBAL BEHAVIOUR

Now that we have some understanding of the way in which a flow or diffeomorphism behaves near a hyperbolic fixed point or periodic orbit we look at the consequences for global behaviour on a manifold M.

In particular, let us consider the case of a point x for which $\phi_t x$ approaches a hyperbolic fixed point p as $t \to +\infty$, so that $\omega(x) = p$. If we take t large enough then $\phi_t x$ will eventually be in $W^s_{\ell oc}(p)$ defined for some suitable neighbourhood of p, or, equivalently, x lies in $\phi_{-t} W^s_{\ell oc}(p)$. Therefore if we take the union of the $\phi_{-t} W^s_{\ell oc}(p)$ over all $t > 0$ (or, indeed, over all $t \in R$) we shall capture precisely those points x for which $\phi_t x \to p$ as $t \to +\infty$, or the global *in-set* for p. This is usually called the global *stable manifold* for p, denoted by $W^s(p)$. Thus $W^s(p) = \bigcup_t \phi_t W^s_{\ell oc}(p)$. It may not be strictly a submanifold of M, since it may wind around the manifold in places rather like the orbits of an irrational flow on a torus, but it can be shown always to be an injectively immersed submanifold (see page 151) - in fact an immersed copy of R^s. It can never 'cross over itself', since iteration of ϕ_t would bring this crossing into some $W^s_{\ell oc}(p)$ which we know to be a genuine submanifold. See Figure 47.

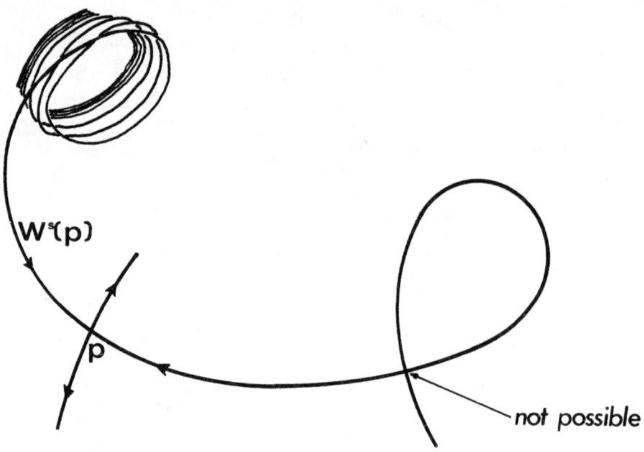

Figure 47

Similarly we define the global *unstable manifold* $W^u(p)$ by

$$W^u(p) = \bigcup_t \phi_t W^u_{\ell oc}(p) ;$$

it consists of all points x with $\phi_t x \to p$ as $t \to -\infty$, i.e. all points which 'come from' p, and is an injectively immersed copy of R^u.

An analogous discussion applies to diffeomorphisms: we define the global *stable* and *unstable manifolds* $W^s(p)$, $W^u(p)$ of the hyperbolic fixed point p to be the result of iterating f infinitely many times forwards and backwards to $W^s_{\ell oc}(p)$, $W^u_{\ell oc}(p)$, and we see that $W^s(p)$, $W^u(p)$ are the global in-set and out-set of p, respectively. They are injectively immersed copies of R^s, R^u. Remember that in both the flow and diffeomorphism cases the dimensions s, u are calculated directly from the eigenvalues of the appropriate linearization at p.

Again we can easily extend these ideas to periodic orbits. If q is a hyperbolic periodic point of f with period m we define $W^s(q)$ by using f^m in place of f and observe that $W^s(f(q)) = fW^s(q)$, i.e. the (global) stable manifold for f^m at fq is precisely the image under f of the (global) stable manifold for f^m at q. The union of the m manifolds, each an immersed copy of R^s,

$$W^s(q),\ W^s(f(q)),\ \ldots,\ W^s(f^{m-1}(q))$$

is called the *stable manifold* for the periodic orbit q.

If γ is a hyperbolic periodic orbit of the flow ϕ then we define $W^s(\gamma)$ by

$$W^s(\gamma) = \bigcup_t \phi_t\, W^s_{\ell oc}(q)$$

where q is a point on γ and $W^s_{\ell oc}(q)$ is defined using a Poincaré map P as before. Here $W^s(\gamma)$ is *not* an immersed copy of a euclidean space, since it is roughly speaking γ's worth of immersed copies of R^s, where s is the number of eigenvalues of $DP(q)$ having modulus less than 1. It can be an injectively immersed copy either of $S^1 \times R^s$ or of $S^1 \tilde{\times} R^s$, which means $[0,1] \times R^s$ with the ends identified by an orientation-reversing diffeomorphism. When $s = 1$ the second case gives a Möbius strip; when $s = 2$ it is a kind of 'solid' Klein bottle (cf. §3.2, [9]).

Replacing f by f^{-1} or ϕ_t by ϕ_{-t} we obtain global *unstable manifolds* for periodic orbits.

Intersections of stable and unstable manifolds

If in the long term a point x comes from p_1 and goes to p_2 then x lies in both $W^u(p_1)$ and $W^s(p_2)$, where by using this notation we are automatically supposing p_1, p_2 to be *hyperbolic* fixed points. Therefore information about the ways in which stable and unstable manifolds (of fixed points or periodic orbits) intersect can be interpreted as information about the global, long-term behaviour of the dynamical system. Here are some observations on diffeomorphisms along these lines:

1. If p_1, p_2 are distinct fixed points then $W^s(p_1)$ cannot meet $W^s(p_2)$. This is because if x lies in $W^s(p_1) \cap W^s(p_2)$ then $f^n(x) \to$ both p_1 and p_2 as $n \to +\infty$, which is nonsense.

2. If $W^s(p_1)$ meets $W^u(p_2)$ where $p_1 \neq p_2$, then it does so in an infinite number of points. This is because $W^s(p_1) \cap W^u(p_2)$ is invariant (as a set) under the action of f, and no point in it is fixed or periodic since such a point could not $\to p_1$ as $n \to +\infty$. Any x in $W^s(p_1) \cap W^u(p_2)$ has $\omega(x) = p_1$, $\alpha(x) = p_2$.

3. Automatically $p \in W^s(p)$ and $p \in W^u(p)$. If there exists some $x \neq p$ with $x \in W^s(p) \cap W^u(p)$ then (cf. 2 above) there is an infinite number of such x's, all satisfying $\omega(x) = \alpha(x) = p$. Such points are called *homoclinic* points, although sometimes this term is reserved for points where $W^s(p)$ and $W^u(p)$ intersect transversally (cf. §4.4). See Figure 48.

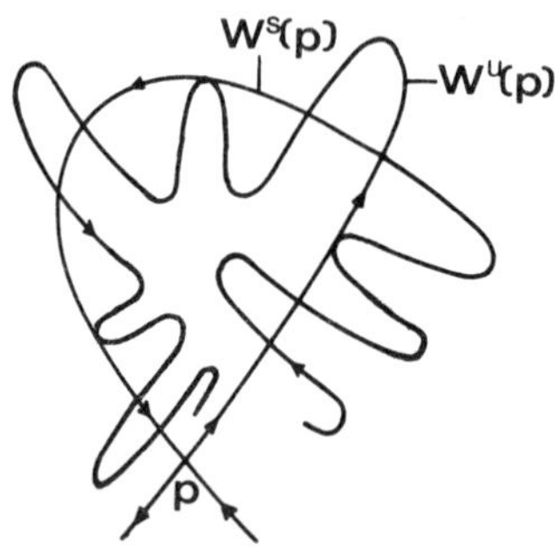

Figure 48

Once again, we can extend these ideas to periodic orbits, and do everything for flows as well as for diffeomorphisms. In this case orbits are no longer discrete, and the behaviour illustrated in Figure 48 could not occur in such low dimensions $(s = u = 1)$. Suspending these diffeomorphism phenomena shows how analogous things happen in higher dimensions for flows.

From the results of this section we see that in studying the global structure of a dynamical system we can make some progress if we know that fixed points and periodic orbits are all *hyperbolic*, and if we know something about the *intersections of stable and unstable manifolds*. Now we might reasonably hope that 'in general' all fixed points and periodic orbits will be hyperbolic, and that stable and unstable manifolds will intersect in general position, i.e. transversally. To discuss this we first have to provide a suitable mathematical meaning for 'in general', and then find which properties we can expect to be those of a 'generic' system. This is the programme for the next section.

4.4 GENERIC PROPERTIES OF FLOWS AND DIFFEOMORPHISMS

It is common practice in classifying any set S of mathematical objects to begin by trying to sort them into two types: *usual* (regular, non-degenerate, tame, nice, ...) and *unusual* (singular, degenerate, wild, pathological, ...). One way to do this is to put a *measure* on the set S, and to associate 'usual' with *occupying a subset of large measure*. Another way is via a topology on S. This is the approach we shall use here, since at the same time it allows us to discuss questions of global stability of dynamical systems.

Naïvely, it seems reasonable to associate 'usual' with *occupying a subset which is dense* in S. This is soon seen to be inadequate, though, since a subset of S and its complement in S can both be dense in S. For example, the subset Q of rational numbers is dense in R, but so is its complement, the set I of irrational numbers, and therefore the rationals and irrationals are both 'usual'. Nevertheless, the irrationals are in some sense 'more usual' than the rationals: there are more of them, for a start. We can capture this by using the topological notion of a *residual set*.

DEFINITION

Let S *be any topological space. A subset* G *of* S *is called a* residual *set if it is the intersection of a* countable *number of sets, each of which is both* open and dense *in* S.

In any respectable topological space a residual set is itself *dense*. This is known as Baire's Theorem, once 'respectable' is defined, and then a space in which this result holds is called a *Baire space* (L. Baire, 1874-1932). For example, any space homeomorphic to a complete metric space is a Baire space. The intersection of a finite or countably infinite number of residual sets is residual, and so in a non-empty Baire space the complement of a residual set cannot be residual (otherwise the empty set would be dense). Therefore *occupying a residual set* becomes a good synonym for 'usual' in a Baire space. Conveniently, the spaces we deal with will all be Baire spaces.

In the rational-irrational example above we see that as the rational numbers are countable we can write $Q = \{q_1, q_2, \ldots, q_n, \ldots\}$ and, writing

U_n for the open dense subset $R - \{q_n\}$ of R, we can then express I as the intersection of all the U_n. Thus I is a residual subset of R, and Q is not.

The sets we are interested in are the set $\mathrm{Diff}^r(M)$ of C^r diffeomorphisms on the smooth n-manifold M, and the set of C^r vector fields on M which we denote by $X^r(M)$. We will now put topologies on these sets. Assume unless otherwise stated that M is *compact*: this makes the topologies easier to describe, and is in any case the context in which our later theorems will apply.

We regard $\mathrm{Diff}^r(M)$ as a subset of the set $C^r(M,M)$ of C^r maps $M \to M$, topologize $C^r(M,M)$, and then take the subset topology induced on $\mathrm{Diff}^r(M)$. To topologize $C^r(M,M)$, or, more generally, $C^r(M,N)$, we want to capture the idea of two C^r maps $M \to N$ being 'close' and we can do this by saying that f, g are close if their local representatives in all charts of some atlases are close. The procedure for converting these hazy notions into a topology is as follows.

Let (U,ϕ), (V,ψ) be charts on M, N with (for technical reasons) the closure $\overline{fU}$ of fU bounded and contained in V. By Theorem C in §1.6 this implies that $f\bar{U}$ is compact. For any $\varepsilon > 0$ let $B_\varepsilon(f;U,V)$ denote the set of all g in $C^r(M,N)$ for which $g\bar{U} \subset V$ and such that

$$\sup_{x \in U} ||\psi f(x) - \psi g(x)|| < \varepsilon \qquad (*)$$

where $||\cdot||$ is any particular norm on R^n. Take atlases for M and N chosen carefully so that there is a cover of M by charts $U_{\alpha_1}, U_{\alpha_2}, \ldots, U_{\alpha_s}$ with the property that each $f\bar{U}_{\alpha_i}$ is bounded and contained in some chart V_{β_i} on N. This is not hard to do, by considering the cover of M by all the open sets of the form $f^{-1}(V_\beta)$, using the compactness of M to get a finite subcover, and then taking slightly smaller charts to satisfy the closure condition. Now define $B_\varepsilon(f)$ to be the intersection of the $B_\varepsilon(f;U_{\alpha_i},V_{\beta_i})$ for $i = 1,2,\dots,s$: then g lies in $B_\varepsilon(f)$ precisely when all its local representatives with respect to the charts U_{α_i}, V_{β_i} differ from those of f by less then ε. Finally, we define a set W in $C^r(M,N)$ to be an *open* set if, given any $f \in W$, there exists some $B_\varepsilon(f)$ constructed as above such that $B_\varepsilon(f)$ is contained in W. This gives a topology on $C^r(M,N)$ the C^0-*topology*, which for N = M induces the C^0-*topology* on $\mathrm{Diff}^r(M)$. The topology can be shown to be independent of the choice of atlases and special charts U_{α_i}, V_{β_i}.

If we want the topology to capture closeness of derivatives of all orders up to r we do the same as above, but replace (*) by

$$\sum_{k=0}^{r} \sup_{y \in U'} \|D^k\bar{f}(y) - D^k\bar{g}(y)\| < \varepsilon , \qquad (**)$$

where $U' = \phi U$, $\bar{f} = \psi f \phi^{-1}$ and (see §2.6) $D^k\bar{f}(y)$ belongs to the space $M_k(R^m \times R^m \times \dots \times R^m;R^n)$ of k-multilinear maps and $\|\cdot\|$ is obtained from norms on R^m and R^n. This then gives us the C^r-*topology* on $C^r(M,N)$, $C^r(M,M)$ and on $\mathrm{Diff}^r(M,M)$. Unless it is stated otherwise, this topology is assumed to be the one in use. It is known more classically as the topology of *uniform convergence of all derivatives up to order* r. The case r = 1 will most interest us.

Remarks

1. There is a more sophisticated way of describing these topologies using *jets*: see Golubitsky and Guillemin [42], for example.

Alternatively we can embed N in some R^p (see §3.2, [6]), and then regard $f : M \to N$ as $f : M \to N \to R^p$. This has an r-fold tangent map $T^r f : T^r M \to T^r R^p = R^q$ where $q = 2^r p$ and with a little care the ordinary euclidean metric on R^q can be used to give a 'distance' between $T^r f$ and $T^r g$. We obtain a metric on $C^r(M,N)$, and this induces the C^r topology already defined. Using the metric, it can be shown that $C^r(M,N)$ and $\mathrm{Diff}^r(M,M)$ are Baire spaces.

2. When M is not compact problems arise since there may not be a finite atlas, and even when there is one a topology constructed as above may not be suitable for dealing with behaviour 'at infinity'. We will not go into details, except to mention that for M, N open subsets of R^m, R^n a useful C^r topology is obtained by replacing the constant ε in (**) by a continuous function $\varepsilon : M \to R$ (which may tail off to zero towards the edge of M). Alternatively, one can consider a topology of *uniform convergence on compact subsets* of derivatives up to order r.

3. Further subtleties occur even in the compact case if we wish to treat all derivatives for C^∞ maps: there are a number of topologies one could use. One choice is to say a set is *open* in $C^\infty(M,N)$ when it is a union of sets, each open in some $C^r(M,N)$.

4. These ideas can be extended to infinite-dimensional manifolds and to manifolds with boundary.

For $X^r(M)$ the procedure is similar but easier. Given a vector field X and a chart $\phi_\alpha : U_\alpha \to U_\alpha'$ there is the local representative $\tau_\alpha X \phi_\alpha^{-1} : U_\alpha' \to U_\alpha' \times R^n$ (cf. p.169) and considering only the second factor we have $X_\alpha : U_\alpha' \to R^n$. If we shrink U_α a little if necessary we can suppose that ϕ_α is defined on the closure $\bar{U}_\alpha$, and if $\phi_\alpha \bar{U}_\alpha$ is bounded in R^n (hence compact) it follows that $||D^k X_\alpha||$ is a bounded function on $\phi_\alpha U_\alpha$. Let $||X_\alpha||_r = \sum_{k=0}^{r} ||D^k X_\alpha||$. When M is compact we can choose a finite atlas of charts $U_{\alpha_1}, U_{\alpha_2}, \ldots, U_{\alpha_s}$ as above, and define

$$||X||_r = \sup_{1 \leq i \leq s} ||X_{\alpha_i}||_r .$$

This is a norm for $X^r(M)$, making it into a Banach space. The norm itself depends very much on the choice of atlas, but the topology induced on $X^r(M)$ does not. (Cf. Example 6 on page 8, where there is only one chart.)

Remarks analogous to those in the $\mathrm{Diff}^r(M)$ case also apply to $X^r(M)$.

With topologies assigned to these spaces of diffeomorphisms and vector fields we can now discuss genericity.

DEFINITION

A property P *of diffeomorphisms (vector fields) on* M *is said to be* C^r generic *if the set* P *of diffeomorphisms (vector fields) possessing* P *contains a residual subset of* $\mathrm{Diff}^r(M)$ $(X^r(M))$.

There are two particular properties P_1, P_2 which are known to be C^1 generic for diffeomorphisms and vector fields on a compact manifold:

P_1 *Every fixed point and periodic orbit is hyperbolic, and all intersections of their stable and unstable manifolds are transversal.*

This does not say that there need be only a finite number of fixed points and periodic orbits. As we shall see below, the property of having only a finite number of periodic orbits is *not* C^1 generic.

P_2 *The non-wandering set* Ω *is precisely the closure of the set of fixed points and periodic orbits.*

Remarks

1. In fact P_1 is C^r generic for each finite r. This is known as the Kupka-Smale Theorem. See Kupka [65], Smale [122] or Peixoto [96].

2. The result that P_2 is C^1 generic is due to Pugh, and follows from his Closing Lemma [103]. This states that any *recurrent point* (a point p such that $p \varepsilon \omega(p)$) can be converted into a periodic point by arbitrarily C^1-small perturbations of the diffeomorphism or vector field. The proof is extremely difficult. For C^0-small perturbations it is easy, and for C^r-small perturbations $(r > 1)$ little is known: the C^r Closing Lemma may be false. It is unknown whether P_2 is C^r generic or not. Thus the choice of topology is not merely a refinement for mathematicians, but has important implications for the kinds of behaviour in dynamical systems that we are to consider as usual or as exceptional.

4.5 GLOBAL STABILITY

We will now take a different point of view from that of genericity, and consider when the overall structure of a dynamical system persists under sufficiently small perturbations of the system. After all, we should be suspicious of a differential or difference equation used to model a real-life system if the equation can be made to behave very differently by including arbitrarily small extra terms, since in practice we can only measure to within some non-zero margin of error. This leads to the idea of *structural stability* of diffeomorphisms and vector fields. We might indulge in some wishful thinking, and hope that structural stability, when suitably formulated, is a generic property ...

Previously (§4.2) we discussed local equivalence of germs of diffeomorphisms or vector fields at a fixed point. We can easily transfer this to a global concept.

DEFINITION

Two diffeomorphisms $f, g \in \text{Diff}^r(M)$ *are* C^k equivalent *(for some* $k \leq r$*) if there exists* $h \in \text{Diff}^k(M)$ *with* $h \cdot f = g \cdot h$.

This is a special case of conjugacy of two maps (§1.5), i.e. f and g 'look the same' globally. Note that in particular h takes f-orbits globally to g-orbits.

DEFINITION

Two vector fields $X, Y \in X^r(M)$ *are* C^k equivalent *if there exists* $h \in \text{Diff}^k(M)$ *taking* X*-orbits to* Y*-orbits, preserving senses but not necessarily parametrization by* t.

In either case h takes the phase portrait of one system to the phase portrait of the other. As pointed out in the local context for fixed points, it is mainly the case $k = 0$ (i.e. h a homeomorphism) that is of practical interest in studying the long-term behaviour of orbits. Therefore our definition of structural stability will be as follows:

DEFINITION

A diffeomorphism $f \in \text{Diff}^r(M)$ *(or vector field* $X \in X^r(M)$*) is* structurally stable *if there is a neighbourhood* N *of* f *in* $\text{Diff}^r(M)$ *(or* X *in* $X^r(M)$*) such that every diffeomorphism (or vector field) in* N *is* C^0 *equivalent to* f *(or* X*).*

If we had worked with C^k equivalence $(k \geq 1)$ we would find that the definition was practically useless. For example, let f be a diffeomorphism with a finite number of fixed points. If f, g are even C^1 equivalent then the conjugating C^1 diffeomorphism h takes each fixed point p of f to a fixed point $h(p)$ of g, and the Chain Rule shows that the linear tangent map $T_{h(p)}g$ is linearly conjugate by $T_p h$ to $T_p f$ and hence has the same eigenvalues as $T_p f$. However, we can easily *change* the eigenvalues of $T_p f$ by arbitrarily small C^1 changes in f, and so f could not be structurally stable. As an example based on the same fact about linearization, the system $\dot{x} = -x$ would not be structurally stable (with $k \geq 1$) since $\dot{x} = -(1+\varepsilon)x$ would not be C^1 equivalent to the first system even though its phase portrait is identical. Of course, if we take $k = 0$ then $T_p h$ does not exist and so neither does the above argument.

Most of the important theorems on structural stability are formulated for compact manifolds. Although the manifolds that it might be necessary to use in practice could often turn out to be non-compact (e.g. phase spaces for second order systems), the results for the compact case show essentially what goes on, and they can often be adjusted to cope with specific non-compact problems. Therefore in what follows we make the standing assumption that M is a C^∞ *compact* n-manifold (without boundary), and we might as well suppose also that M is *connected*, otherwise we could treat each piece of M separately.

The early results on structural stability (called 'roughness') were those of the Russian school (Andronov, Pontrjagin [7]) in the 1930's. However, the first main global theorem which together with Smale's results on gradient systems (see §4.6) launched a new attack on the subject, was Peixoto's theorem for flows on 2-manifolds.

THEOREM (Peixoto [95])

A C^r *vector field on a 2-dimensional manifold* M *is structurally stable if and only if it satisfies the following conditions:*

(1) The number of fixed points and periodic orbits is finite, and each is hyperbolic;

(2) there are no points x *for which* $\alpha(x)$ *and* $\omega(x)$ *are both saddle points;*

(3) Ω *consists of fixed points and periodic orbits only.*

Moreover, if M *is orientable the set of such vector fields is open and dense (thus highly generic) in* $\mathcal{X}^r(M)$.

Remarks

1. As stated in [95] the full result includes the non-orientable case, but there is a gap in the proof. For $r = 1$ the Closing Lemma comes to the rescue, but it seems that the situation has still not been clarified for $r > 1$.

2. Condition (2) is the same as transversality of intersection of stable and unstable manifolds: the only dimensions for which transversality is not automatic are $s = u = 1$ (i.e. for saddle points), and then by uniqueness of solutions if $W^s(p_1)$ is tangent to $W^u(p_2)$ they must meet along a whole orbit. Thus genericity of (hyperbolicity + (2)) corresponds to the Kupka-Smale Theorem (page 227).

With this result in mind, Smale proposed a closer study of systems on compact n-manifolds satisfying the above conditions (1) and (3) together with

(2) All stable and unstable manifolds intersect transversally.

Such systems are now known as *Morse-Smale* (MS) systems. Apart from their dynamic properties, Morse-Smale systems have the added interest of providing close analogues to the Morse inequalities (page 160). There is a set of *Morse-Smale inequalities* (Smale [119], Markus [73]), through which the topology of M places constraints on the possible numbers of fixed points and periodic orbits for any MS diffeomorphism or flow on M. However, from the point of view of dynamics MS systems seemed good material for two plausible conjectures:

Conjecture 1 A system is structurally stable if and only if it is MS.

Conjecture 2 MS systems are dense in $\text{Diff}^1(M)$ or $X^1(M)$.

Failing Conjecture 2, one could at least try

Conjecture 3 Structurally stable systems are dense in $\text{Diff}^1(M)$, $X^1(M)$.

These conjectures were short-lived. Their fates are quickly summarized:

Conjecture 1 MS implies structurally stable: proved TRUE by Palis and Smale (see [22]).
Structurally stable implies MS : FALSE in general.

Conjecture 2 FALSE in general.

Conjecture 3 FALSE in general.

Of course, the conjectures are true for vector fields when $\dim M = 2$ by Peixoto's Theorem. They are also true for diffeomorphisms when $\dim M = 1$, (i.e. $M = S^1$). The failure of Conjectures 1 and 2 in general is a consequence of the fact that there exist complicated non-wandering phenomena which systems can exhibit and which cannot be removed by small perturbations. We will now look at two basic examples of such phenomena. The failure of Conjecture 3 is more subtle, and we shall be in a better position to understand it once we have understood these examples.

EXAMPLE 1: The Smale Horseshoe

Take a rectangle R in the plane R^2. Pick it up, stretch it and bend it into a horseshoe shape as shown (vertices A, B, C, D go to A', B', C', D') and put it down again partly on top of its original position: see Figure 49. It is not difficult, using smooth but possibly non-analytic

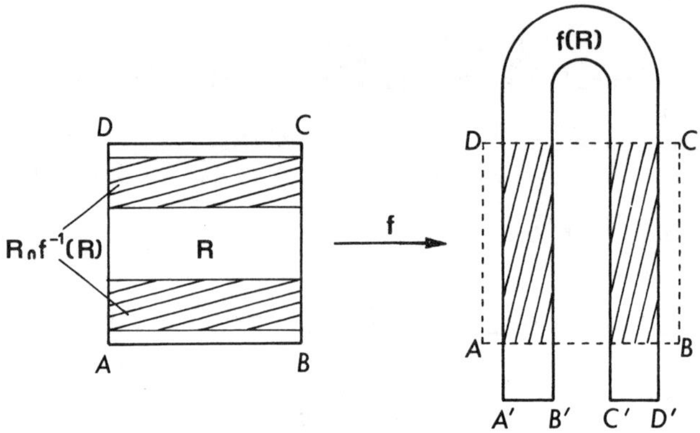

Figure 49

functions (cf. example on page 86) to piece things together, to incorporate this construction into a diffeomorphism $f : R^2 \to R^2$. Furthermore (and this is important) we can arrange that the two rectangular strips $R \cap f^{-1}(R)$ consisting of points of R which remain in R when f is applied to them are each stretched linearly by a factor > 1 in the 'vertical' direction (parallel to AD) and compressed linearly by a factor < 1 in the 'horizontal' direction (parallel to AB).

Now look at $R \cap f^{-1}(R) \cap f^{-2}(R)$, i.e. the set of points of R which remain in R when both f and f^2 are applied to them. This consists of *four* thin rectangles, two in each rectangle of $R \cap f^{-1}(R)$.

Similarly, $R_{-n} = R \cap f^{-1}(R) \cap \ldots \cap f^{-n}(R)$ consists of 2^n thin rectangles contained pairwise in the 2^{n-1} rectangles of $R_{-(n-1)}$. See Figure 50.

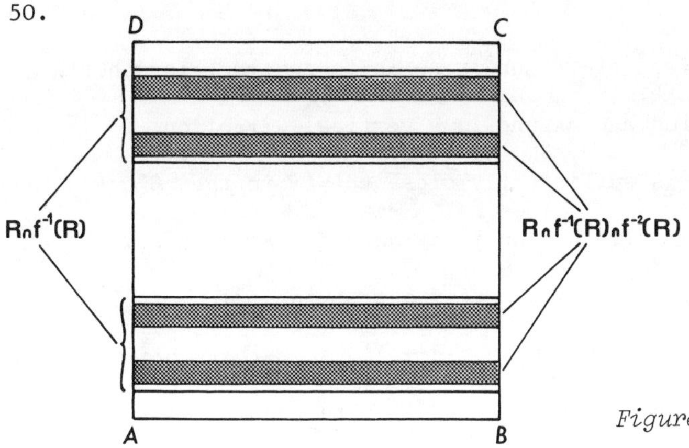

Figure 50

The set

$$R_{-\infty} = R \cap f^{-1}R \cap f^{-2}R \cap \ldots ,$$

which consists of *those points of* R *which remain in* R *after any number of positive iterations of* f, is the cartesian product $I \times C$ where I is a closed interval (corresponding to the width AB of R) and C is an example of a *Cantor set* (G. Cantor, 1845-1918), a closed, uncountable set with empty interior and yet such that every point in C is an accumulation point of C (see e.g. Willard [150] for a discussion of Cantor sets). Likewise, the set R_∞ consisting of *those points of* R *which remain in* R *under all negative iterations of* f is of the form $C \times I$. If Λ denotes $R_{-\infty} \cap R_\infty$ then we see that Λ is a closed *invariant set* for f, meaning that f takes Λ onto itself, and Λ is homeomorphic to $C \times C$, which can be shown to be in fact homeomorphic to C. Some further facts about Λ are these:

(1) There is an infinite number of periodic points of f in Λ. The set of all periodic points in dense in Λ,

so every point of Λ is a non-wandering point for f.

(2) There are points which have orbits that are dense in Λ.

(3) The effect of f near Λ is to contract in a horizontal direction and expand in a vertical direction.

(4) If g is sufficiently close to f in the C^1 topology then g also has an invariant set Λ' satisfying (1) and (2) and approximately satisfying (3). In fact there is a homeomorphism $h : \Lambda \to \Lambda'$ such that $h \cdot f = g \cdot h$ on Λ, so f on Λ and g on Λ' are topologically conjugate.

Facts (1) and (2) are quite easy to prove. Label the two rectangles of $R_{-1} = R \cap f^{-1}R$ by the symbols 0, 1. We can then label the four rectangles of R_{-2} by symbols 00, 01, 10, 11, where e.g. $x \in 01$ when $x \in 0$ and $f(x) \in 1$. Hence we can label each horizontal line in $R_{-\infty}$ by an infinite sequence of 0's and 1's. The same applies to R_∞, and so if we write the R_∞ sequences backwards and adjoin them to the $R_{-\infty}$ sequences we can label each point of Λ by a doubly-infinite sequence of 0's and 1's, provided we keep track of the join, of course. Explicitly, we denote an element x of Λ by

$$\ldots a_{-3}\, a_{-2}\, a_{-1}\, a_0\, a_1\, a_2\, a_3 \ldots$$

where each a_n is 0 or 1, and the sequence indicates that $f^n(x)$ lies in the rectangle a_n of R_{-1} for each integer n. Now the action of f on Λ clearly corresponds to a *shift* of each sequence one place to the left. In this way we can recognize periodic points simply as those whose symbol-sequence is periodic, and once the topology of Λ is expressed as a suitable topology on the set of all symbol-sequences the verification of

Facts (1), (2) becomes routine. This method is an example of the use of *symbolic dynamics*.

Fact (3) is an immediate consequence of the construction of f on R. However, Fact (4), which contains the meat of this example, is not at all easy to prove. It needs techniques for dealing with 'global' perturbations, and these techniques only became available with the work of Hirsch, Pugh, Shub and others following the ideas of Smale in the mid 1960's. See Hirsch, Palis, Pugh and Shub [54], or Nitecki [90].

The horseshoe construction can be incorporated into $f : S^2 \to S^2$ rather than $f : R^2 \to R^2$, and so Fact (4) shows that $\text{Diff}^1(S^2)$ contains an open set consisting of diffeomorphisms which are not Morse-Smale. Hence the failure of Conjecture 2 for diffeomorphisms. By taking suspensions we obtain counterexamples to the conjecture for flows in dimension 3.

EXAMPLE 2: Hyperbolic toral automorphisms

Let A be any 2×2 matrix with integer entries and with determinant equal to $+1$. Suppose also that the eigenvalues of A are not of modulus 1. A standard example is $A = \begin{pmatrix} 1 & 1 \\ 1 & 2 \end{pmatrix}$, which has eigenvalues $\lambda = \frac{3}{2} \pm \frac{\sqrt{5}}{2}$. Such an A represents a linear isomorphism of R^2 to itself; in particular it is a diffeomorphism which has the origin as a hyperbolic fixed point. The stable and unstable manifolds of this point are the straight lines $x_2 = \frac{1}{2}(1 \pm \sqrt{5})x_1$, i.e. the eigenspaces corresponding to the two eignevalues.

Now we use A to define a diffeomorphism of a torus $T^2 = S^1 \times S^1$ to itself as follows. Parametrize S^1 by $e^{2\pi it}$ $(t \in R)$ and define $f : T^2 \to T^2$ by

$$f(e^{2\pi is}, e^{2\pi it}) = (e^{2\pi is'}, e^{2\pi it'})$$

where (s',t') is the image of (s,t) under A. This does make sense because A has integer entries (so altering s and t by integers will at most alter s', t' by integers), and f is a bijection since $\det A = \pm 1$ and thus A^{-1} also has integer entries and induces an inverse to f on T^2. Equivalently, think of T^2 as the quotient space R^2/Z^2 (see page 144) and observe that since A is invertible and preserves Z^2 it automatically induces a bijection $f : T^2 \to T^2$. Clearly f is a C^∞ diffeomorphism, since smoothness is a local property and f looks locally like A which is linear and so C^∞ on R^2. It is called a hyperbolic toral automorphism, where the word *automorphism* is used to emphasize that f preserves the co-ordinate-wise multiplicative structure of $S^1 \times S^1$ as well as the smooth structure. Facts of interest to us about f are the following:

(1) There is an infinite number of periodic points of f. The set of periodic points is dense in T^2, and so the non-wandering set $\Omega(f)$ is the whole of T^2.

(2) There are points which have orbits that are dense in T^2.

(3) The effect of f is to contract in one direction (parallel to $t = \frac{1}{2}(1 - \sqrt{5})s$) and expand in another ($t = \frac{1}{2}(1 + \sqrt{5})s$).

(4) The diffeomorphism f is structurally stable.

Fact (1) is easy to prove, since for each m, f permutes the set of points whose co-ordinates are both rational numbers with denominator m. Fact (2) requires more work, but is straightforward. Fact (3) comes directly from the hyperbolic nature of A on R^2. Fact (4) is again the difficult one, needing global perturbation techniques, and is of fundamental importance. It is a special case of a more general theorem which we discuss below. This gives another example of the failure of Conjecture 2, and also illustrates how Conjecture 1 fails. As it happens, we could have employed the horseshoe example to show this, too, but it would require some further elaboration.

The construction generalizes to $T^n = S^1 \times S^1 \times \ldots \times S^1$, thought of also as a quotient space R^n/Z^n. We obtain corresponding flow versions by taking suspensions.

In view of these examples it seems clear that the idea of *contracting in one direction and expanding in another* in a global sense is somehow related to structural stability, which is not altogether surprizing since we know this is the case locally for hyperbolic fixed points and periodic orbits (§4.2). To measure contraction and expansion globally in general we are going to need to look at local linearizations everywhere, or at least on the parts of the manifold that interest us, and then have some way of comparing the effects of linearization at different places on the manifold. This means that we must consider the action of the tangent map Tf on the tangent bundle TM, and then be able to compare lengths of vectors in different tangent spaces T_xM. We can certainly do this if we have a norm $||\cdot||_x$ on each T_xM, and, as we saw at the end of §3.4, we have this once we have a Riemannian structure on M. Therefore we shall now assume that we have a given Riemannian structure on M, and consequently a norm on each tangent space.

Let Λ denote any closed invariant set for the diffeomorphism f on the compact manifold M. We make the following definition.

DEFINITION

The invariant set Λ *is* hyperbolic *for* f *(or* f *is* hyperbolic *on* Λ*) if for each* x *in* Λ *the tangent space* T_xM *splits into the direct sum of two linear subspaces* E_x^s, E_x^u *such that*

(1) *the splitting varies continuously over* Λ *and is invariant under the action of* Tf, *i.e.*

$T_xf(E_x^s) = E_{f(x)}^s$ *and similarly for* E_x^u, *and*

(2) *there are positive numbers* $\lambda < 1$ *and* c, C *such that*

$$||T_xf^n(v)||_{f^n(x)} < C\lambda^n||v||_x \quad \textit{for all} \quad v \in E_x^s$$

$$||T_xf^n(v)||_{f^n(x)} > c\lambda^{-n}||v||_x \quad \textit{for all} \quad v \in E_x^u$$

for every positive integer n.

Remarks

1. We will avoid defining precisely what 'various continuously' means: intuitively it makes sense. Note that the dimensions s, u need not be constant on all of Λ, but as integer-valued continuous functions they will be constant on each *connected* subset of Λ (see §1.7).

2. The flow version is obtained by replacing f^n by ϕ_t with $t > 0$, and λ^n by λ^t or $e^{\alpha t}$ where $\lambda = e^{\alpha}$.

3. Hyperbolic fixed points and periodic orbits are hyperbolic invariant sets in the above sense.

4. The horseshoe example is hyperbolic on Λ; the toral automorphism example is hyperbolic on the whole torus.

If x is a point of Λ and Λ is hyperbolic for f we would like to define the *(generalized) stable manifold* $W^s(x)$ for x to be the set of all y in M for which the distance between $f^n(y)$ and $f^n(x)$ approaches zero as $n \to \infty$. To do this we must have a metric on M, but since we already have a Riemannian structure we can use that to provide a metric: again see §3.4. Alternatively, we could embed M in euclidean space. Therefore the definition of $W^s(x)$ makes sense (it is independent of the particular metric chosen) and there is then the Generalized Stable Manifold Theorem, due to Hirsch and Pugh [53], which we summarize:

THEOREM

Each $W^s(x)$ *is an immersed copy of* R^s, *with* E^s_x *as tangent space at* x. *As* x *runs through* Λ *all the* $W^s(x)$ *fit together to give a continuously-varying family (at least locally near* Λ), *and the family is invariant in that* $W^s(f(x)) = fW^s(x)$.

Replacing n by $-n$ we obtain the definition and analogous theorem for generalized unstable manifolds $W^u(x)$.

This theorem in the case $\Lambda = \Omega(f)$ is one of the main ingredients in the long-awaited Structural Stability Theorem that generalizes Peixoto's theorem and the theorem of Palis and Smale on MS systems:

THEOREM

Suppose the diffeomorphism f *satisfies the following:*

(1) the non-wandering set Ω *is hyperbolic;*

(2) periodic orbits are dense in Ω;

(3) all intersections of (generalized) stable and unstable manifolds are transversal.

Then f *is structurally stable.*

Remarks

1. This theorem was proved by Robbin [105] as the culmination of work by many people including Hirsch, Pugh, Shub and Smale: see [22], [115]. In Robbin's proof it is necessary to suppose that f itself is actually C^2, although small perturbations are allowed in $\mathrm{Diff}^1(M)$. A proof in the C^1 case was given by Robinson [108].

2. The converse (that structural stability implies (1), (2) and (3)) is almost proved, and is certainly true if the definition of structural stability is strengthened a little: see Guckenheimer [46], Franks [39]. In fact Franks [40] shows that (1), (2), (3) are necessary and sufficient conditions for a slightly amended structural stability property that may be of more relevance from a practical point of view than the previous 'pure' definition.

3. Corresponding results exist for flows. See Robinson [107], Franks [38].

Behaviour of structurally stable systems

We will here use the word *system* to mean either a diffeomorphism or a flow.

The two conditions on the non-wandering set Ω

$$\left.\begin{array}{l} \Omega \ \textit{hyperbolic} \\ \textit{periodic points dense in } \Omega \end{array}\right\}$$

together comprise what Smale [125] called *Axiom A*. Thus Robbin's structural stability theorem can be stated as

$$\left.\begin{array}{l}\textit{Axiom A plus transversality of}\\ \textit{stable and unstable manifolds}\end{array}\right\}\text{ implies } \textit{structural stability}\ .$$

No examples are known of systems with Ω hyperbolic but periodic points *not* dense in Ω, and so it is an unsolved problem to decide whether the first clause of Axiom A always implies the second.

The structure of Ω for an Axiom A system can be described in fairly general but useful terms via Smale's *Ω-decomposition theorem* [125]:

THEOREM

If Axiom A is satisfied then Ω *breaks up into a disjoint union*

$$\Omega = \Omega_1 \cup \Omega_2 \cup \ldots \cup \Omega_s$$

in which each Ω_i *is* closed, invariant *under the diffeomorphism (or flow), and such that the diffeomorphism (or flow) acts* transitively *on each* Ω_i, *i.e. has an orbit dense in* Ω_i.

It is possible that $s = 1$ and $\Omega = \Omega_1$ is the whole of the manifold M: this was the case for the hyperbolic toral automorphism. In general, systems for which the whole manifold is hyperbolic are called *Anosov* systems, since they first came to light in the flow context through work of the Russian D.V. Anosov [8]. It is not known whether hyperbolicity of M always implies $\Omega = M$. The classification of Anosov systems is an interesting and very difficult open problem, important because an Anosov system with $\Omega = M$ is an example of a dynamical system that exhibits complicated recurrent behaviour everywhere which cannot (since it is structurally stable) be destroyed by small C^1 perturbations.

The sets Ω_i in the decomposition are called *basic sets*. They generalize hyperbolic fixed points and periodic orbits as models for stable (thus observable) recurrence phenomena in dynamical systems.

The Cantor set in the horseshoe is an example of a complicated basic set. In principle we would like to be able to classify all possible types of basic sets Ω_i for Axiom A systems, and especially those which are *attractors*, i.e. for which $W^s(\Omega_i)$ is a neighbourhood of Ω_i (so that everything near Ω_i approaches Ω_i as t (or n) $\to +\infty$). These generalize *sinks*, which are hyperbolic fixed points p with $\dim W^s(p) = \dim M$. Some interesting results have been obtained by Williams [151].

There is a school of thought that strange attractors in smooth dynamical systems may be the right models for hydrodynamic turbulence. See Ruelle and Takens [110], Iooss [60], Marsden et al. [75], Arnol'd [10].

The behaviour of Axiom A diffeomorphisms and flows on the basic sets Ω_i themselves has been studied in detail by Bowen, who showed that symbolic dynamics can be used to give a very good picture of the measure - theoretical or statistical properties of the system. The idea of entropy plays an important role. Thus deterministic mechanics leads via Axiom A to statistical mechanics within the same model. See Bowen [19].

To understand the global behaviour of an Axiom A system we have to see which points x go from one Ω_i to another Ω_j, i.e. satisfy $\alpha(x) \subset \Omega_i$ and $\omega(x) \subset \Omega_j$. Such an x lies in the intersection of $W^u(\Omega_i)$ and $W^s(\Omega_j)$. (This is more difficult to prove than it looks. It amounts to showing that if $d(f^n(x), \Omega_j) \to 0$ as $n \to +\infty$ then there is some point y in Ω_j with $d(f^n(x), f^n(y)) \to 0$ as $n \to +\infty$, i.e. $x \in W^s(y)$. See Figure 51. The proof uses strongly the density of periodic points in Ω, and the result is false for an arbitrary hyperbolic invariant set Λ : see the counterexample due to Bowen in [54].)

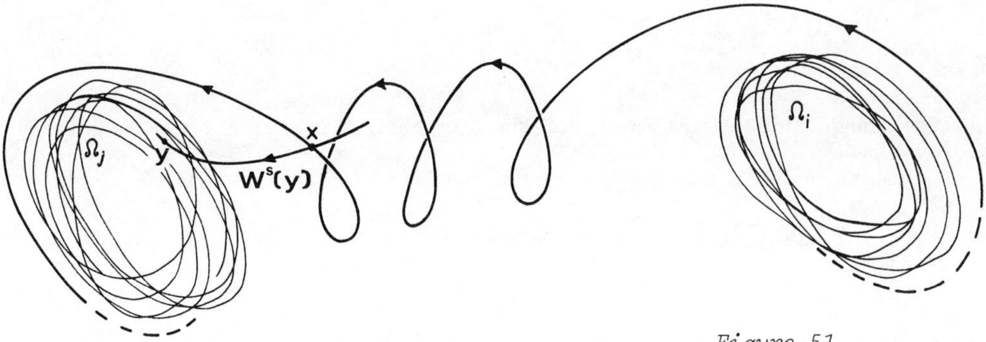

Figure 51

Putting an ordering on the Ω_i's by writing $\Omega_i > \Omega_j$ to mean that $W^u(\Omega_i) \cap W^s(\Omega_j)$ is nonempty, we can draw a schematic diagram or graph with vertices representing the Ω_i's and a directed edge joining Ω_i to Ω_j whenever $\Omega_i > \Omega_j$. It can be shown that the condition of transversal intersection of stable and unstable manifolds implies that there are no cycles (closed circuits of consistently directed edges) in the graph, so that the global behaviour of the system is represented by a hierarchical structure as in Figure 52. The attractors are the Ω_i's at the bottom

Ω_5, Ω_8 are attractors

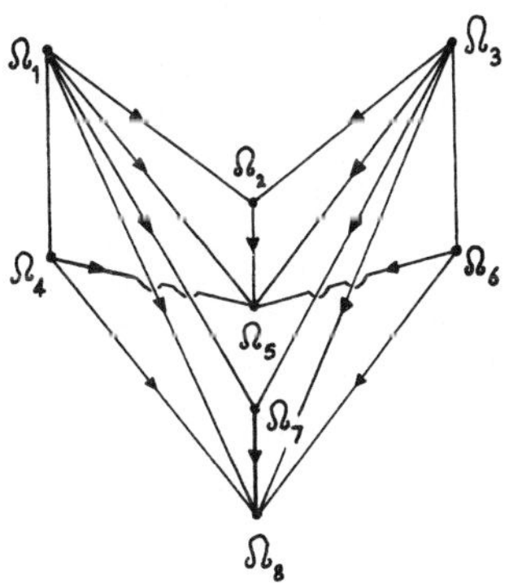

Figure 52

of the hierarchy. In general it is not known which graphs can come from Axiom A systems, or to what extent equivalent graphs imply equivalent systems, although a classification in the case of Morse-Smale systems on 2-manifolds has been carried out by Peixoto [98]

Final remarks on stability

1. The example due to Smale [124], showing that structural stability is unfortunately not a generic property in $\mathrm{Diff}^1(M)$, was obtained by constructing an Axiom A diffeomorphism f_0 having an ordinary saddle point p as one basic set Ω_1 and a torus with a hyperbolic toral automorphism on it as another basic set Ω_2, and then changing f_0 to f_1 for which $W^u(p)$ bends by a large amount and has non-transversal intersection with $W^s(y)$ for some $y \in \Omega_2$. If g is sufficiently C^1 close to f_1 this non-transversality persists, with p', y', Ω_2' for g close to p, y, Ω_2 for f_1. The type of phase-portrait for g depends on whether or not y' is periodic or non-periodic in Ω_2', and both possibilities occur densely. This means that f_1 has a C^1 neighbourhood of diffeomorphisms, none of which is structurally stable. Hence structurally stable systems are not dense in $\mathrm{Diff}^1(M)$, and Conjecture 3 is demolished. The same holds for vector fields, by taking a suspension.

2. Despite this result, it turns out that every system in $\mathrm{Diff}^1(M)$ or $\mathcal{X}^1(M)$ can be approximated arbitrarily C^0-closely by structurally stable systems: see Shub [116], Zeeman [152]. This is a rather confusing result to interpret, since the definition of structurally stable uses C^1 perturbations and yet implies C^0 density. The significance of this depends on what choices of topologies seem appropriate in any given context of mathematical modelling.

3. There are other global notions of stability that have been proposed in the hope that they may be both physically meaningful and (in some useful topology) generic. For example, *Ω-stability* (stability of the system restricted to Ω : see e.g. Shub [115]) is not C^1 generic; *tolerance stability* (due to Zeeman; see Takens [137]) is in a certain sense generic.

4.6 DYNAMICAL SYSTEMS UNDER CONSTRAINT

The results considered above about structural stability and genericity all assume that arbitrary perturbations in $\mathrm{Diff}^r(M)$ or $X^r(M)$ are to be allowed, whereas there are many circumstances in modelling real systems in which the admissible class of perturbations will be restricted and so possibly lead to entirely different results both about stability and genericity. In fact systems under constraint are probably more common than unconstrained systems. We will consider three types of constrained system: *gradient systems*, *Hamiltonian systems* and *systems with symmetry* and comment on their global stability properties.

To discuss gradient and Hamiltonian systems we need to introduce the *cotangent bundle* of a smooth manifold.

Cotangent bundles

Let $f : M \to R$ be a smooth function on a manifold M modelled on a Banach space E. For each point p on M we have the linear tangent map

$$T_p f : T_p M \to T_{f(p)} R = R$$

and $T_p f$ is an element of $(T_p M)^* = L(T_p M, R)$. Elements of $(T_p M)^*$ are called *covectors* at p, and $(T_p M)^*$ - which we rewrite as $T_p^* M$ - is the *cotangent space* of M at p. Usually we denote T_pf by $df(p)$: then df assigns a covector to each point p of M, and is thus an example of a *covector field* on M.

To discuss the way $df(p)$ varies with p we need to put a topology, and preferably a smooth structure, on the set T^*M of all covectors at all points on M. This can be done by analogy with the construction for the tangent bundle (§3.4). Choose a chart $\phi : U \to U'$ with $p \in U$, identify T_pM with E (as on page 167) and hence identify T_p^*M with $E^* = L(E,R)$. Corresponding to the chart

$$\tau : TU \to U' \times E$$

for TM we obtain a map

$$\tau^* : T^*U \to U' \times E^*$$

and so an atlas for M gives us an atlas for T^*M as a manifold modelled on the cartesian product $E \times E^*$. If the atlas for M is C^r then the atlas obtained for T^*M is C^{r-1}. The proof of this is similar to that for TM, although a little more subtle. Whereas the TM proof relied on the fact that a C^r overlap map $f : U \to V$ induces a C^{r-1} map $U \times E \to V \times F$ taking (p,h) to $(f(p),Ah)$ where $A = Df(p) \in L(E,F)$ (equation (**) on page 165), this time we have to use the fact that the diffeomorphism $f : U \to V$ induces a C^{r-1} map $U \times E^* \to V \times F^*$ taking (p,h) to $(f(p), A^{-1*}h)$, where for $L \in L(F,E)$ the map $L^* \in L(E^*,F^*)$ takes $\mu \in E^*$ to $\mu \cdot L \in F^*$.

Thus T^*M is a C^{r-1} differentiable manifold called the *cotangent bundle* for M. As with TM we have a natural projection map $\pi : T^*M \to M$, and a covector field on M is a section of the bundle, i.e. a map $\alpha : M \to T^*M$ satisfying $\pi . \alpha = id_M$. Covector fields are also called 1-*forms* or *Pfaffians* (J. Pfaff, 1765-1825).

In finite dimensions we often regard R^{n*} as R^n by blurring the distinction between $1 \times n$ matrices and $n \times 1$ column vectors. We are really saying that since (given a choice of coordinates) any linear map $\lambda : R^n \to R$ can be expressed as $x \mapsto \ell_1 x_1 + \ell_2 x_2 + \ldots + \ell_n x_n = \langle \ell, x \rangle$ we will identify $\lambda \in R^{n*}$ with $\ell = (\ell_1, \ell_2, \ldots, \ell_n) \in R^n$. More briefly, we let $\lambda(\cdot)$ correspond to $\langle \ell, \cdot \rangle$ where $\langle \cdot, \cdot \rangle$ is the standard inner product (page 52) on R^n. This works equally well without coordinates on any finite-dimensional linear space V if we replace the inner product by any non-degenerate bilinear map

$$\beta : V \times V \to R \quad .$$

There is a map $\beta_\flat : V \to V^*$ taking v to $\beta(v, \cdot)$; the non-degeneracy implies that this is injective, and then by counting dimensions it must also be surjective, i.e. an isomorphism. In the case of a Hilbert space (page 53) the properties of the inner product are such that we again get an isomorphism between H and H^*. This fact was used in discussing critical points of functions: see page 107.

Suppose that for each point p on the smooth manifold M we have a non-degenerate bilinear map

$$\beta(p) : T_pM \times T_pM \to R.$$

Then we have $\beta_\flat(p) : T_pM \to T_p^*M$, and in the finite-dimensional or Hilbert space case we also have

$$\beta_\#(p) = \beta_\flat(p)^{-1} : T_p^*M \to T_pM.$$

Hence we can apply $\beta_\#(p)$ to convert any covector at p to a vector at p, and so β converts any 1-form into a vector field.

To discuss continuity or smoothness of this process we construct a bundle over M with fibre $M_2(T_pM \times T_pM ; R)$ over p, obtaining a smooth manifold called the *tensor bundle of covariant degree* 2. We give no details, but the analogy with T^*M is close: see e.g. Abraham [1]. A section β of this bundle is called a *covariant tensor field of degree* 2, and is *non-degenerate* if the bilinear map $\beta(p)$ is non-degenerate for every p in M. All that we have said so far is summarized by: *any smooth covariant tensor field* β *of degree* 2 *on* M *converts a smooth covector field* α *into a smooth vector field* X, *namely the unique vector field satisfying*

$$\beta(p)\ (X(p),v) = \alpha(p)v$$

for all $v \in T_pM$ *and every* p *in* M.

Now we look at two particular cases.

Gradient systems

If $\beta(p)$ is symmetric and positive definite for every p then β is a *Riemannian structure* on M (see page 178). Given a C^r function $f : M \to R$ the covector field df is converted by β into a C^{r-1} vector field called the *gradient field* of f with respect to β, denoted by $\text{grad}_\beta f$ or just $\text{grad}\ f$ is β is understood. For example, when $M = R^n$ and $\beta(p)$ is the standard inner product for all p then $df(p)$ is the $1 \times n$ matrix

$$Df(p) = \left(\frac{\partial f}{\partial x_1}(p),\ \frac{\partial f}{\partial x_2}(p),\ldots,\frac{\partial f}{\partial x_n}(p)\right)$$

and so $\text{grad}\ f$ is the same expression regarded as a field of vectors in R^n. This was our definition of gradient system in §4.1 (Example 4). If we had chosen a different inner product of the form $\langle x,y\rangle = \sum_{i,j=1}^{n} g_{ij}x_iy_j$ (with g_{ij} possibly depending on p) then $df(p)$ would not change but the ith component of the gradient vector at p would become $\sum_{j=1}^{n} h_{ji}\frac{\partial f}{\partial x_j}(p)$ where (h_{ij}) is the matrix inverse to (g_{ij}).

A gradient field X on a (compact) manifold M gives rise to a flow ϕ with particularly simple properties. If we write p_t for $\phi_t(p)$ we have

$$\begin{aligned}\frac{dt}{dt} f(p_t) &= df(p_t) \cdot \dot{\phi}_t(p) \\ &= df(p_t) \cdot X(p_t) \\ &= \beta(p_t)\,(X(p_t), X(p_t))\end{aligned}$$

by definition of X, and this last term is $\geqslant 0$ since $\beta(p_t)$ is positive definite. Hence f increases along the orbits of X, and so there can be *no periodic or recurrent phenomena*. The non-wandering set Ω consists of fixed points alone, namely the critical points of f. Moreover, the non-fixed orbits of ϕ are perpendicular (in terms of β) to the level sets f = constant (remember (page 153) that away from critical points they are *manifolds*) since if $Y(p)$ is a tangent vector to $N = f^{-1}(k)$ then $\beta(p)(X(p),Y(p)) = df(p)Y(p) = 0$. This gives a very good grasp of the behaviour of gradient systems, and Smale [120] was able to prove that *gradient systems with hyperbolic fixed points and transversal intersections of stable and unstable manifolds are structurally stable* as one of the first main results of the theory. Of course, we now see this as a special case of the general structural stability theorem in §4.5. Interestingly, Smale's Ω-decomposition theorem and the graph representation (page 243) allow any Axiom A system with transversality of intersection of stable and unstable manifolds to be regarded in a very loose way as a kind of overall gradient system.

Hamiltonian systems

If $\beta(p)$ is anti-symmetric (i.e. $\beta(p)(u,v) = -\beta(p)(v,u)$ for every u,v in T_pM) then β is called a 2-*form* on M. As above, a non-degenerate

2-form will convert any covector field into a vector field. However, this time we see that df becomes a vector field X which is not perpendicular to the level sets (manifolds away from critical points of f, = zeros of X) but is tangent to them, since we have

$$\frac{d}{df}f(p_t) = \beta(p_t)(X(p_t),X(p_t)) = 0$$

because of the anti-symmetry, and so f remains constant along orbits. This means that f represents some quantity which is *conserved* by the flow. For most k the level set $f^{-1}(k)$ is a codimension-1 submanifold of M (recall Sard's Theorem, page 157), and so for fixed such k the phase space for X has been reduced to dimension n-1. If for some reason we had to apply to the system a constraint of the form $f(x) \leqslant k$ we would then be working with a flow on a manifold with boundary (cf. §3.2, [8]). If β satisfies a further technical condition (namely, if it is *closed*: see e.g. Abraham [1] or Spivak [128]) then it becomes a *symplectic* form, and the vector field is called a *Hamiltonian* vector field (W.R. Hamilton, 1805-1865) with f in this context called a *Hamiltonian* function.

It so happens that the cotangent bundle T*N of any manifold N has a natural symplectic 2-form that can be defined on it, and so any smooth (Hamiltonian) function $f : T^*N \to R$ gives automatically a Hamiltonian vector field X on T*N. (Remember that here $M = T^*N$, so we have $X : T^*N \to T(T^*N)$.) If we take N to be an open set U in R^n then it turns out that the vector field X on $T^*U = U \times R^{n*} = U \times R^n$ corresponds to the system of ordinary differential equations

$$\dot{x}_i = \frac{\partial f}{\partial y_i}, \qquad \dot{y}_i = -\frac{\partial f}{\partial x_i} \qquad i = 1,2,\ldots,n$$

where (x,y) are the coordinates in $U \times R^n$. These are *Hamilton's equations* from classical mechanics.

The global behaviour of Hamiltonian flows is very complicated, since it turns out that there can be no sources or sinks, and recurrence phenomena abound. There are some quite simple properties known to be generic for Hamiltonian flows (see Robinson [106], Takens [135]), and also some extraordinarily elaborate ones (Markus and Meyer [74]). In view of these, it seems difficult to know how to formulate global stability conjectures within the Hamiltonian context. Of course, the characteristically Hamiltonian behaviour can always be destroyed by allowing perturbations through non-Hamiltonian systems in $X^1(M)$. A common example is the simple harmonic system (Example 1, §4.1) in which the family of periodic orbits around the origin is converted into a spiral configuration with the origin as a sink by the addition of arbitrarily small amounts of frictional damping. Here the undamped system is Hamiltonian with $f(x_1,x_2) = k^2x_1^{\ 2} + x_2^{\ 2}$ and symplectic 2-form $(u,v) \mapsto u_2v_1 - u_1v_2$ for $u,v \in R^2 = T_pR^2 = T_p(T^*R)$. The damped system is not Hamiltonian.

Systems with symmetry

If a dynamical system is subject to some symmetry constraint then stability and genericity results will only be relevant inasmuch as they apply to the world of diffeomorphisms or flows constrained by the same symmetry. The proper setting for this is *equivariant* differential topology, where manifolds, maps and vector fields are all studied in the context of an action of a Lie group G which has to be respected at all times. It is an important and largely open problem to express and then prove meaningful genericity and stability theorems for G-equivariant diffeomorphisms and flows, both in general and for specific types of group G. Some important basic results have been obtained by Field [37].

4.7 BREAKDOWN OF STABILITY: BIFURCATION THEORY

Structurally stable diffeomorphisms (or vector fields) form an open set S in $\mathrm{Diff}^r(M)$ (or $X^r(M)$). We have previously considered the behaviour of elements of S, but now let us look at elements not in S. Restricting attention to the vector field case, we denote the complement of S in $X^r(M)$ by Σ, for some given r understood. The set Σ is called the *bifurcation set* in $X^r(M)$. It represents places where the phase portrait of the flow changes in topological type as X moves around in $X^r(M)$: these changes are called *bifurcations*.

If X lies in Σ then we may or may not be able to perturb X by arbitrarily small amounts in the C^1 topology to elements Y which are in S (if $\dim M = 2$ then Peixoto's theorem (page 230) says that we can; Smale's example (page 244) and others show that if $\dim M > 2$ it may not be possible). Let us assume X to be such that we can perturb it in this way to make it structurally stable, and consider the following question.

Suppose X_λ, $0 \leqslant \lambda \leqslant 1$, is a smoothly-varying family of vector fields on M such that $X_{\lambda_0} = X$ for some particular λ_0 with $0 < \lambda_0 < 1$ and such that X_λ is structurally stable for $\lambda \neq \lambda_0$. In other words, we have a smooth path

$$c : [0,1] \to X^r(M)$$

which meets Σ in X and otherwise lies in S. (Recall that $X^r(M)$ is a Banach space (page 226), so there is no problem about defining any amount of smoothness for c.) Can we perturb the entire path c by arbitrarily small amounts to make it avoid Σ ? If locally near X the bifurcation set Σ contains a kind of codimension-1 hypersurface in $X^r(M)$, a wall separating X_0 from X_1 , then the answer will be No. See Figure 53.

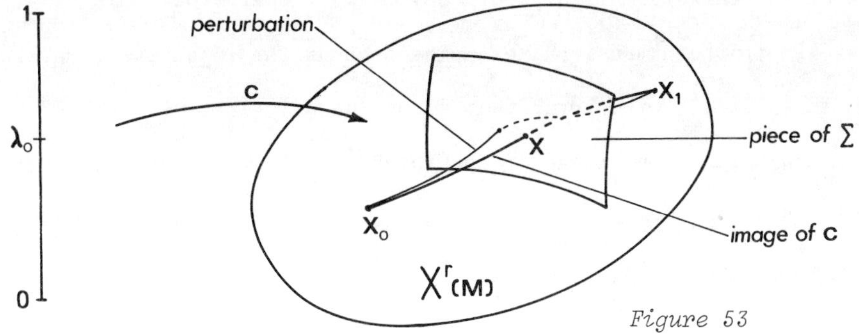

Figure 53

Similarly, we could consider k-parameter families of vector fields and try to classify types of element of Σ by whether or not they can be avoided by k-parameter families in general. The larger the k needed to 'catch' a certain type of element of Σ , the rarer the type of instability. This kind of approach was originated by Thom, and has led to much extremely interesting mathematics, both in the study of dynamical systems and elsewhere. We will pursue these ideas in two particular directions.

Recall from 53.2 [6] that if M is a smooth infinite-dimensional manifold we say that a submanifold W is of *codimension* k if every point p on W has an open neighbourhood U in M such that $W \cap U = f^{-1}(0)$ for some submersion $f : U \to R^k$ with $f(p) = 0$. This implies that M is diffeomorphic to $W \times R^k$ locally near p , and so k is the smallest dimension of manifold that can meet **W** transversally. Any piece of ℓ-dimensional manifold in U would in general not meet W if $\ell < k$. This agrees with the definition of codimension as dim M − dim W in the case where M is finite-dimensional. As a slight extension to the definition, we say that W is an *immersed submanifold of codimension* k in M if $W = \bigcup_{i=1}^{\infty} W_i$ where each W_i is a submanifold of codimension k and $W_i \subset W_{i+1}$ for each i, or in other words W is the union of an increasing family of submanifolds all of codimension k. Observe how this is already the case in finite dimensional examples such as the injective immersions of $\mathbb{R}$ in $S^1 \times S^1$ with irrational slope (page 151) : here we can take W_i to be the image of the interval $(-i,i)$, a genuine submanifold of $S^1 \times S^1$. The increasing family is a device for copying this kind of intuition in infinite dimensions.

The codimension measures the 'avoidability' of W. This was used by Sotomayor [126] to characterize a large part of Σ corresponding to vector fields on 2-manifolds which only just fail to be structurally stable.

THEOREM

The set Σ_1 *of vector fields on a 2-manifold whose flow satisfies all the conditions of Peixoto's theorem with just one infringement (and that in a certain sense as non-degenerate as possible) is a codimension-1 immersed* C^{r-1} *submanifold of* $X^r(M)$ *, at least for* $r \geq 4$*. Moreover,* Σ_1 *is dense in* Σ *.*

As a consequence of the theorem, Sotomayor shows that generically a 1-parameter family of vector fields X_λ will meet Σ_1 transversally, and this then allows the changes of behaviour of the flow as X_λ passes through Σ to be described quite fully. The picture is not quite as simple as in Figure 53 however, not merely because Σ_1 is only an immersed submanifold of $X^r(M)$, but because there are parts of Σ not on Σ_1 but still unavoidable by 1-parameter families. This seems difficult to imagine, but is related to the complicated topological structure of Σ that corresponds to rational and irrational flows on a torus. Possibly if the torus were excluded then $\sum_1$ would represent the only unavoidable bifurcations occurring in 1-parameter families of vector fields on 2-manifolds.

These global bifurcation results do not extend easily to flows in n dimensions, because of the complications caused by (suspended) horseshoes and similar pheonomena. However, there are some interesting theorems about 1-parameter families: see Newhouse [88], Newhouse and Palis [89]. *Local* bifurcations at fixed points or periodic orbits can in many cases be understood just as well as those in two dimensions: see Sotomayor [127] for a discussion of codimension-1 behaviour. The problem of analyzing codimension-k bifurcations at fixed points of flows for $k > 1$ has been attacked by Takens [136], and although he gives a classification for $k \leqslant 3$ he shows that this 'codimension' approach becomes impractical for $k \geqslant 4$ in general.

The importance of bifurcation theory in the mathematical modelling of real systems is pointed out very strongly by Thom [138], who has set up the framework for a very general and all-embracing theory of development of form or *morphogenesis* which in principle could apply to biology, psychology and linguistics as well as to more physical sciences. The theory is based on the idea of codimension as sketched above, and although the results of Takens made it clear that the technicalities of the mathematics may be more difficult than originally hoped, there is an important special class of system for which (as Thom showed) the mathematical machinery does work exceedingly well. Let us say that a flow ϕ on M is *governed by a potential function* f if $f : M \to R$ is a smooth function with the properties that

(1) the fixed points of ϕ are precisely the critical points of f;

(2) $f(\phi_t(p))$ is a decreasing function of t, for every point p in M other than the fixed points of ϕ.

The flow from $X = -\text{grad } f$ would satisfy these conditions, although there are plenty of other examples where f is some kind of energy function. For such a flow the non-wandering set consists of the critical points of f, with the stable equilibria corresponding to minima. If we suppose that the flow is so rapid that transient effects die out quickly and the system is only ever observed essentially in an equilibrium state, then the bifurcation theory has reduced to the theory of coalescences and (dis)appearances of critical points of smooth functions f. Now it is a consequence of the Morse lemma (page 104) that *a non-degenerate critical point of* f *will persist under all sufficiently* C^2*-small perturbations of* f, and therefore bifurcations can only occur at degenerate critical points. Since this is a local problem, we may as well assume $M = R^n$. Thom's theorem can be paraphrased as follows:

THEOREM

There are only seven types of bifurcations of (degenerate) critical points which cannot in general be avoided by k*-parameter families of smooth functions, where* $k \leq 4$.

Here *type* means equivalence class under suitable local change of coordinates: see the further discussion below. The important thing to note about the theorem is that the number of types depends only on k and *not on the number* n *of variables*. Bifurcations in general, or the changes of phase-portrait that they represent, are called *catastrophes* by Thom. The local bifurcations of critical points, or the germs at the critical points themselves, are *elementary catastrophes*.

An immediate objection to the use of this theorem is that systems are rarely governed by potential functions, as such systems can have no recurrent or even periodic behaviour. This is so, but nevertheless the bifurcations of periodic orbits or more complicated basic sets can often be reduced to the potential function case by artificial mathematical devices such as generalized Liapunov functions (A.M. Liapunov, 1857-1918): see Shub [115]; see also Hirsch and Smale [55] or Arnol'd [11] for classical Liapunov theory at fixed points of flows.

Towards a more precise formulation of Thom's theorem

As in §2.7, let E_n denote the set of germs at 0 of C^∞ functions $R^n \to R$. For convenience of writing we will be rather lax about distinguishing between germs and functions that represent them. We have previously said that two such germs f,g are *right equivalent* if there exists a germ of a diffeomorphism $\gamma : R^n \to R^n$ taking 0 to 0 and such that $g = f \cdot \gamma$. Since we are interested only in critical points, we will now broaden the definition a little, and say that f,g are *equivalent* if $g = \pm f \cdot \gamma + K$ for some constant K. Suppose that we have a topology on E_n. Then we can define an element f of E_n to be *stable* if it has some neighbourhood N in E_n for which every g in N is equivalent to f. Following the general approach outlined at the beginning of this section, we let $\sum$ denote the subset of E_n consisting of those which are not stable. To study $\sum$ we try to analyze the 'avoidability' of various pieces of it in general k-parameter families of maps $R^n \to R$, for $k = 1,2,\ldots$.

The programme behind Thom's Theorem is as follows. We ought to be able to regard E_n as an infinite-dimensional manifold. Any element with non-vanishing first derivative at 0 is stable (since it is right equivalent to its derivative (see page 110) and any two non zero linear maps $R^n \to R$ are right equivalent), and so Σ lies in the codimension-n submanifold Σ_0 consisting of germs with vanishing derivative. Further, Σ ought to be decomposible as a collection of disjoint (although abutting) pieces $\Sigma_1, \Sigma_2, \ldots$ where each Σ_i is a codimension-i submanifold of Σ_0, together with a very 'small' remaining piece Σ_∞. Each Σ_i should consist of a finite number of classifiable equivalence classes. Given a k-parameter family of maps $f_c : R^n \to R$ (where $c = (c_1, c_2, \ldots, c_k) \in R^k$) we define a *global germ map*

$$F : R^n \times R^k \to E_n$$

by $F(x,c)$ = germ of $y \mapsto f_c(x+y)$ at $y = 0$. In other words, $F(x,c)$ is the germ at 0 of f_c with origin moved to x. In general we hope for this map to be transversal (page 176) to all the Σ_i, and therefore to meet only those Σ_i with $n+i \leqslant n+k$, avoiding Σ_∞ as well. Finally, two k-parameter families whose global germ maps meet a given Σ_i transversally at the same point should themselves be in some sense locally equivalent.

The initial obstacle to carrying out this programme is that E_n cannot easily be made into a suitable manifold. A way around this, though, is to convert the problem into a finite-dimensional one by working with jets of some given order, look in this linear space of jets (see page 88) for submanifolds representing the Σ_i, and then to use facts about determinacy of jets (§2.9) to lift the results back up to the germ level. A second problem is that the decomposition of Σ only works in the simple way suggested for $i \leqslant 5$: after that, natural candidates for Σ_i tend to contain infinite families of equivalence classes. However, Thom was interested in $k \leqslant 4$ parameters for reasons mentioned below, and so this problem is avoided.

The genericity of transversality of F to the $\sum_i$ is expected as a consequence of Thom's powerful Transversality Theorem:

THEOREM

If M,W *are manifolds and* S *is a closed submanifold of* M *then the set of maps* $W \to M$ *which are transversal to* S *is an open dense subset of* $C^\infty(W,M)$.

This theorem has been the springboard for many important advances in differential topology. In fact we need here a more sophisticated version to show (a) perturbations in f_c supply enough perturbations of F to obtain transversality with $\sum_i$, and (b) the fact that the submanifolds $\sum_i$ may not be closed does not matter since together they form a *stratification*. The required version exists, also due to Thom.

By writing $\Phi(x,c)$ instead of $f_c(x)$ we may as well regard a k-parameter family locally as a germ $\mathrm{R}^n \times \mathrm{R}^k \to \mathrm{R}$ at $(0,0)$, which we will assume to be smooth in x and c and therefore an element of E_{n+k}. We call Φ an *unfolding* of the germ $f = f_0$. Two unfoldings Φ,Ψ of f are *equivalent* if there exist smooth germs

$w : \mathrm{R}^n \times \mathrm{R}^k \to \mathrm{R}^n$ with $x \mapsto w(x,c)$ a diffeomorphism germ for each c, and $w(x,0) = x$

$b : \mathrm{R}^k \to \mathrm{R}^k$ a diffeomorphism germ with $b(0) = 0$

$a : \mathrm{R}^k \to \mathrm{R}$ with $a(0) = 0$

such that

$$\Psi(x,c) = \Phi(w(x,c),b(c)) + a(c) . \qquad (*)$$

Thus Ψ is obtained from Φ by (i) an invertible change of x-variables, depending on c, (ii) an invertible change b of c-variables and (iii) addition of a number a depending on c. We call the unfolding Φ *stable* if, roughly, any other unfolding (of f) sufficiently C^∞-close to Φ is equivalent to Φ. In trying to make this precise we again run up against the problem of a topology for E_{n+k}, but one way around this is to choose a specific neighbourhood U of (0,0) to work with, using a C^∞ topology for maps $U \to R$, and then to insist on stability on U for every U. See Wassermann [146] for a very careful discussion of these questions.

Now we can give a more technical expression of Thom's theorem.

THEOREM

For any $n > 1$ *there are seven equivalence classes* $\Sigma_1^1, \Sigma_2^1, \Sigma_3^{1,2,3}, \Sigma_4^{1,2}$ *(called* strata*) in* E_n *such that*

(1) *each* Σ_i^j *is a submanifold of* E_n *of codimension* $n+i$;

(2) *generically, for a* k*-parameter family* Φ *with* $k \leq 4$ *the corresponding global germ map* F *will be transversal to the* Σ_i^j *and will avoid the rest of* Σ ;

(3) *such a* Φ *is* versal, *meaning that if the origin is chosen in* R^k *such that* Φ *is an unfolding of* $f \in \Sigma_i^j$ *then any other unfolding* Ψ *of* f *can be obtained from* Φ *by coordinate changes as in* (*) *except that* b *need not be invertible. If* $k = i$ *then* Φ *is called* universal : *every other universal unfolding of* f *is equivalent to* Φ *. From this and the definition of transversal it follows that universal unfoldings are stable.*

Remarks

1. The interpretation of the manifold structures is strictly speaking via jets, as mentioned above.

2. For $n = 1$ there is only one stratum of each codimension: see the comments on Thom's classification list below.

3. For $k = 5$ we include four more strata $\Sigma_5^{1,2,3,4}$, but for $k \geqslant 6$ we would need to include an infinite family and this would complicate the genericity property (2).

4. Under our definition f is equivalent to $-f$. A sign change converts minima (= stable equilibria) into maxima and *vice-versa*, so in applications it is important to distinguish between f and $-f$. If we do this the seven strata break into ten (not 14, since e.g. x^3 can be changed to $-x^3$ by $x \mapsto -x$).

5. We supposed Φ defined on all of $R^n \times R^k$ only for convenience; any open subset would do for the domain of Φ.

6. Full details of the machinery for proving the theorem were essentially published in more general form in the work of Mather [78]. The main tool for dealing with unfoldings on the jet level is the Preparation Theorem, a classical theorem for formal power series only recently proved by Malgrange for C^∞ functions. See references at the end of this chapter.

A universal unfolding of f gives a full description of the way the degenerate critical point at 0 breaks up into a cluster of critical points nearby when f is perturbed in *any way whatsoever* that is small in the appropriate C^∞ sense, thus reducing a seemingly infinitely complicated problem into a finite geometrical one. For any unfolding Φ, the set K of points $c \in R^k$ for which f_c has at least one *degenerate* critical point is the *catastrophe set* K. As c passes through K a bifurcation occurs for a flow on R^n governed by f_c. For Φ, Ψ universal unfoldings of f the diffeomorphism b in (*) takes the catastrophe set for Φ onto that for Ψ, and so the catastrophe set for any particular universal unfolding is a kind of archetype representing the catastrophe sets for all other universal unfoldings.

Thom has given a now celebrated list of seven germs representing each of the equivalence classes $\sum_i^j$, together with universal unfoldings for them and (theoretically) pictures of the corresponding archetypal catastrophe sets. We will not give the list, since it may be found in any of the references to catastrophe theory given below: perhaps this is the first account of the theory in which the list does not appear! Four germs are of the form $f(x) + Q$ while three are of the form $f(x,y) + Q$ where Q is a non-degenerate quadratic form in the remaining n-1 or n-2 variables. Recall that from the Gromoll-Meyer lemma (page 107) any germ at a degenerate critical point can be put into the form *(totally degenerate) + (non-degenerate)*. Implicit in Thom's list is the theorem that any germ whose degenerate part needs more than two variables will be avoidable in generic k-parameter families, $k \leqslant 4$.

Given an unfolding of a germ with degenerate critical point at 0 it is important to know how to find out whether the unfolding is universal or not. There is a formal criterion for this: *the unfolding* $\Phi : R^n \times R^k \to R$ *of* f *is universal if and only if, given any germ* $g \in E_n$, *there exist germs* $h_i \in E_n$ $(1 \leqslant i \leqslant n)$ *and numbers* a_j $(0 \leqslant j \leqslant k)$ *such that*

$$g(x) = a_o + \sum_{j=1}^{k} a_j \frac{\partial \Phi}{\partial c_j}(x,0) + \sum_{i=1}^{n} h_i(x) d_i(x)$$

where $d_i(x) = \frac{\partial f}{\partial x_i}(x)$ *as in* §2.9.

EXAMPLES

1. $n = 1$; $f(x) = x^4$. Here $\Phi(x;c_1,c_2) \equiv x^4 + c_1x + c_2x^2$ is a universal unfolding since any g can be written as

$$g(x) = a_0 + a_1 \frac{\partial\Phi}{\partial c_1} + a_2 \frac{\partial\Phi}{\partial c_2} + 4x^3h$$

$$= a_0 + a_1x + a_2x^2 + 4x^3h$$

for some germ h. This is the *cusp* catastrophe.

2. $n = 2$; $f(x,y) = x^3 + y^3$. The unfolding

$$\Phi(x,y) \equiv x^3 + y^3 + c_1x + c_2y + c_3xy$$

is universal, since any g can be written as

$$g(x,y) = a_0 + a_1 \frac{\partial\Phi}{\partial c_1} + a_2 \frac{\partial\Phi}{\partial c_2} + a_3 \frac{\partial\Phi}{\partial c_3} + 3x^2h + 3y^2k$$

for some germs h,k. This is the *hyperbolic umbilic* catastrophe.

In general when Φ is not a polynomial there will be no finite way of checking this criterion directly, so we need to replace the problem by one about polynomials rather as in the tests for determinancy in §2.9. This can be done easily enough, although it is a little elaborate to write down. See Poston and Stewart [101].

The use of catastrophe theory in practice

To interpret the significance of the catastrophe set in a given family of systems we have to know how the critical points of the function f_c are observed, via the physical or other laws which operate in the particular context. For example, when there are two or more minima of f_c available the system may adopt the state corresponding to the lowest minimum: in this case the set of points c for which f_c has two equal-valued minima (the *shock-wave set*) will play a crucial role, rather than the catastrophe set itself.

Thom's use of catastrophes to study morphogenesis in biology is based on the following model. If T is a piece of biological tissue in space and time we think of T as a region in R^4. Each cell C in T can be regarded as a point $c = (c_1, c_2, c_3, c_4) \varepsilon R^4$, and then if we hypothesize that the internal behaviour of C is described by a dynamical system governed by a potential function f_c we have a 4-parameter family Φ and can apply the foregoing theory. The catastrophe set and shock-wave set are subsets of T itself, and through the laws of adhesion and other properties of cells they will influence the observed manifestations of the bifurcations in f_c. Since we are modelling a real-life system we can assume that Φ will be everywhere locally stable as an unfolding (an assumption vital to the model) and hence since there are $k \leq 4$ parameters the local forms for Φ and the catastrophe set K will up to local coordinate change be among those on the list of seven.

We now outline three general programmes for using catastrophe theory in applications. For a much more detailed analysis of catastrophe modelling see the discussion between Thom and Zeeman [140], and also Zeeman [155].

1. Contemplate the seven catastrophe sets and their associated shock-waves, and then look at the real world to try to find phenomena which exhibit geometric properties analogous to them. Conjecture that these phenomena can be modelled by systems governed by potential functions, and after selecting the appropriate archetypes from Thom's list persuade the biologist, physicist etc. to search for the corresponding small numbers of key variables and parameters which will exhibit these phenomena as catastrophes.

(Thom [138], [139]; Zeeman [154]).

2. Observe which bifurcations occur in existing mathematical models, and check whether the unfoldings are stable. (Unstable unfoldings will be detected if they give catastrophe sets not equivalent to any on Thom's list.) If not, then improve the model by including extra parameters to make the unfoldings stable. (Berry [14], Chillingworth and Furness [28]).

3. Use the qualitative features of a real-life system and/or a mathematical abstraction to suggest the existence of a particular underlying catastrophe. Then use the geometry of the catastrophe set to explain other qualitative features, preferably including some which may have seemed difficult to explain otherwise. (M. Dodson [34]; Sewell [113], Thompson and Hunt [142]; M. Thompson [143]; Zeeman [153], [157].)

Further extensions of the theory

Apart from the application of Thom's methods to bifurcations of general flows and diffeomorphisms that cannot be reduced to the straight catastrophe formalism, some of the main directions in which developments are likely to occur are:

1. *Catastrophe theory with symmetry.* One substantial objection to catastrophe theory is that many systems are constrained by symmetry and/or conservation laws, and so the genericity and stability assumptions in catastrophe theory fall to the ground: cf. §4.6. Indeed they do, although this simply means that a theory must be developed that operates within these constraints. The basic machinery for this has now been set up by Poénaru [100], and soon there should be 'Thom lists' available under a variety of different types of symmetry hypothesis. For a discussion of bifurcations of Hamiltonian systems see Meyer [161].

2. *Use of higher codimensions.* Although a generic k-parameter family of functions will avoid all of $\sum$ except the $\sum_i$ with codimension $\leqslant k$ in $\sum_0$, the rest of $\sum$ will still make itself felt in the way the $\sum_i$ fit together and hence produce coherent clusters of elementary catastrophes. The behaviour of germs at critical points is now well understood up to quite high codimensions: see Arnol'd [12]. Critical point germs of high codimension occur when low codimension bifurcations are coupled together. An important example of this studied so far is the *double cusp* catastrophe $x^4 + y^4$: see Poston, Stewart and Woodcock [102], Zeeman [158].

3. *General singularity theory.* The machinery of stability, genericity and unfoldings works for singularities of maps $R^n \to R^m$ for cases other than $m = 1$. Singularities which persist (up to local equivalence) under small perturbations are *stable*. Stable singularities $R^2 \to R^2$ were classified by Whitney [149]: they have local form either $(x_1, x_2) \mapsto (x_1, x_2^2)$ (a *fold*) or $(x_1, x_2) \mapsto (x_1, x_2^3 + x_1x_2)$ (a *cusp* as in Example 4, page 102). Much is now known about stability of singularities and unfoldings of unstable ones, following the work of Mather [78]. See Golubitsky and Guillemin [42]. This information should soon find its way into any sphere of mathematical modelling concerned with m functions of n variables, as it has done already in mathematical economics.

4. *Infinite dimensions.* In continuum mechanics bifurcation theory has developed many techniques of its own (Keller and Antman [63], Sattinger [112]) and is only now beginning to interact with catastrophe theory. Sometimes the problems can be reduced to finite-dimensional ones, using the Gromoll-Meyer lemma (page 107) as in Chillingworth [27], or the powerful Centre Manifold Theorem as in Marsden, Ebin and Fisher [75] or Marsden and McCracken [76]. The ideas of genericity can sometimes be

used directly in the appropriate function-space: see Chow, Hale and Mallet-Paret [29]. There are also ways in which catastrophe theory can be used in partial differential equations: see Guckenheimer [47], Duistermaat [35]. Since partial differential equations are ultimately of more use in modelling the real world than are ordinary differential equations, it is probably here that catastrophe theory will in the long run have its greatest impact.

Remarks on the literature. Most of the recent work in global stability and genericity for dynamical systems stems from the survey of Smale [125]. See in addition the lecture notes by Markus [73], and the books by Nitecki [90] and Abraham and Robbin [2]. The progress of research can be traced through the three Proceedings volumes [22], [97], [71], each of which also contains useful survey articles. For a treatment of differential equations with accessible proofs of some of the key results used in this chapter see the forthcoming book by Irwin and Robertson [62]. The monograph by Moser [85] relating classical and new theories in the context of celestial mechanics is highly recommended reading. In this connection see also the appealing book on differential topology and dynamical systems by Abraham [1].

The main reference in catastrophe theory is of course the treatise of Thom [138]. For mathematical background see Bröcker and Lander [21], Poston and Stewart [101], Wassermann [146], Zeeman and Trotman [159], and also the excellent survey of bifurcation theory by Arnol'd [10]. There are numerous expository articles about catastrophe theory. A very elementary introduction is Chillingworth [26]; for treatments with more mathematical content see Golubitsky [41], Sussmann [133]. The general article by Stewart [132] is valuable reading, although at present there is nothing to surpass the *tour de force* of Zeeman [156].

And if I've said too much, they'l say;

I'm Sorry not at all;

For much more unto Such, I may,

And not be Criminall.

Thomas Mace : *Musick's Monument*, 1676.

Appendix

TERMINOLOGY AND NOTATION FOR SETS AND FUNCTIONS

Sets

1. A *set* is a collection of objects specified either by a given list or by some defining property or process. (There are logical paradoxes inherent in this definition, but we keep well clear of them.) The objects in a set are called its *elements* : we write $a \in A$ to denote that a is an element of A or *belongs* to A. In many contexts elements are considered as *points*. The set consisting of elements $a_1, a_2, \ldots$ is denoted by $\{a_1, a_2, \ldots\}$, and if A is defined as the set of elements b of some other set B satisfying a certain property P we write $A = \{b \in B \mid b \text{ satisfies } P\}$. Standard sets we will use are:

$\mathbb{N}$ = natural numbers $\{1,2,3,\ldots\}$

$\mathbb{Z}$ = integers $\{0,\pm1,\pm2,\ldots\}$

$\mathbb{Q}$ = rational numbers $\{p/q$ where $p,q \in \mathbb{Z}$ and $q \neq 0\}$

$\mathbb{R}$ = real numbers

$\mathbb{C}$ = complex numbers.

We will not define $\mathbb{R}$, $\mathbb{C}$ here, but we assume familiarity with them.

2. The *cartesian product* $A \times B$ is the set of all pairs (a,b) where the first object a is an element of A and the second object b is an element of B. Examples are the *euclidean plane* $\mathbb{R}^2 = \mathbb{R} \times \mathbb{R}$ and, more generally, *euclidean n-space* defined inductively as $\mathbb{R}^n = \mathbb{R}^{n-1} \times \mathbb{R}$.

3. In a particular case the symbol (a,b) may also be used to mean the *open interval* from a to b, i.e. the set of real numbers x with

$a < x < b$ denoted formally by $(a,b) = \{x \in R \mid a < x < b\}$. The *closed interval* $[a,b]$ is $\{x \in R \mid a \leqslant x \leqslant b\}$. There are also (with the obvious meaning) *half-open* intervals $(a,b]$ and $[a,b)$, as well as *infinite* intervals (a,∞), $(-\infty,b)$, and $(-\infty,\infty) = R$.

4. If every element of A is automatically an element of B then A is a *subset* of B or is *contained in* B : we write $A \subset B$. This allows the possibility $A = B$, although some authors prefer the notation $A \subseteq B$ if equality is allowed. Note that *contained in* does not mean *belongs to*, which applies to elements and not subsets. If $A \subset B$, the set of everything in B which does *not* belong to A is the *complement* of A in B, written B-A when there is no possible confusion with subtraction in an algebraic sense. The *union* $A \cup B$ of A and B is the set consisting of everything in either A or B ; the *intersection* $A \cap B$ is the set of everything in both A and B. It is easy to see that if A and B are subsets of C then

$$C - (A \cup B) = (C - A) \cap (C - B)$$

$$C - (A \cap B) = (C - A) \cup (C - B) .$$

It is convenient to think formally of an *empty set* $\emptyset$ which has no elements at all, so that for example if A,B have no elements in common we can write $A \cap B = \emptyset$. Here we say A and B are *disjoint*.

5. A relation R between pairs of elements of a set A is called an *equivalence relation* if it is *reflexive* (aRa for every $a \in A$), *symmetric* (aRb implies bRa for every $a,b \in A$) and *transitive* (aRb and bRc together imply aRc for every $a,b,c \in A$). For given a, the set $\{x \in A \mid xRa\}$ is called the *equivalence class* of a with respect to R : it contains a itself in view of reflexivity. It follows from symmetry and transitivity that any two equivalence classes either coincide or are

disjoint.

6. If A is a subset of $\mathbb{R}$ a number M is called an *upper bound* for A if $a \leq M$ for all $a \in A$. If M_0 is an upper bound for A and there exists no smaller M which is also an upper bound for A, then M_0 is a *least upper bound* or *supremum* for A: we write $M_0 = \sup A$. Similarly for *lower bound* and *greatest lower bound* or *infimum*, written $\inf A$.
It is either a theorem or an assumption about real numbers (depending on your definition of $\mathbb{R}$) that every subset of $\mathbb{R}$ with an upper bound has a least upper bound; similarly for lower bounds.

Functions

7. A *function* f from A to B is a prescription for associating precisely one element of B to each element of A. We write

$$f : A \to B$$

and denote the effect of f on a by $f(a)$. Sometimes the brackets are omitted for simplicity. If $f(a) = b$ we write $f : a \mapsto b$, to be read as "f takes a to b". In formulae it is occasionally convenient to indicate f as $f(\cdot)$, with the dot keeping track of where the 'variable' a has to be inserted in the formula. The word *map* is generally synonymous with *function*, although it is customary in analysis and topology to call a map a function only in the case when $B = \mathbb{R}$.

The prescription for $f(a)$ must be defined for *every* element a of A, or we are not strictly entitled to put A at the beginning of the arrow. The set A is the *domain* of the map f. If U is a subset of A, we may wish to consider the operation of f on elements of U alone. This *restriction* of f to U is written $f|U$. In full, this is $f|U : U \to B$.

8. The set B is the *codomain* of f, a rather formal definition since

the codomain could be changed by including B in some larger set and thinking of this as the codomain without essentially doing anything to f. Not every element of the codomain need come via f from anything in A. The subset of B consisting of those elements that *do* come from something in A is called the *range* of f, or the *image* of f, or the *image of* A *under* f, in any case written fA. If fA is the whole of B, so that everything in B comes from something in A, then f is *surjective* (is a *surjection*).

9. Two distinct elements of A may well be taken by f to the same element of B. If this never happens, then f is *injective* (is an *injection*). Thus f is injective if and only if

$$f(a) = f(a') \quad \text{implies} \quad a = a' .$$

10. If f is both injective and surjective it is *bijective* (is a *bijection*). In other words, it is 'one-to-one and onto'. The *identity map* $id_A : A \to A : a \mapsto a$ is a bijection. A bijection $f : A \to B$ has an *inverse* $f^{-1} : B \to A$ defined by

$$f^{-1}(b) = \text{unique } a \in A \text{ satisfying } f(a) = b .$$

11. If $f : A \to B$ and $g : B \to C$ are any two maps then they have a *composition* $g \cdot f : A \to C$ defined as $a \mapsto g(f(a))$ for every $a \in A$. Note that for this to make sense the domain B of g must be the same as the codomain of f, although the image of f could be smaller than B. The inverse f^{-1} of a bijection $f : A \to B$ is characterized by the relationships

$$f^{-1} \cdot f = id_A : A \to A$$

$$f \cdot f^{-1} = id_B : B \to B .$$

12. Even if f is not a bijection, we can still define for each subset V of B a subset of A called the *inverse image* of V under f, consisting

of those elements a of A for which $f(a)$ belongs to V. This inverse image is denoted by $f^{-1}(V)$, so

$$f^{-1}(V) = \{a \in A \mid f(a) \in V\} .$$

Remember, however, that the symbol f^{-1} on its own is meaningless unless f is bijective. (In this case the two meanings of $f^{-1}(V)$ coincide.) If V consists of just one element v, we write $f^{-1}(v)$ instead of the more cumbersome but formally correct $f^{-1}(\{v\})$.

EXAMPLES

(i) $f : \mathsf{R} \to \mathsf{R} : x \mapsto \sin x$ Neither injective nor surjective.
$f\mathsf{R}$ = closed interval $[-1,1]$. $f^{-1}(0) = \{n\pi \mid n = 0,\pm1,\pm2,\ldots\}$.

(ii) $f : \left[-\frac{\pi}{2},\frac{\pi}{2}\right] \to [-1,1] : x \mapsto \sin x$ Bijective.

(iii) Let M_n = $n \times n$ matrices (real numbers as entries), $n \geq 1$. Define $d : M_n \to \mathsf{R}$ by $d(M) = \det(M)$. Then d is surjective but not injective. $d^{-1}(\mathsf{R} - \{0\})$ is the set of non-singular matrices.

(iv) Let S be the unit sphere in R^3, and let θ,ϕ be the Euler angles (see almost any book on mechanics). Define $f : (0,\pi) \times (0,2\pi) \to S$ by $f(\theta,\phi)$ = point on S with Euler angles θ,ϕ. Then f is injective but not surjective. If instead we define $F : [0,\pi] \times [0,2\pi] \to S$ by the same wording then F is surjective but not injective.

13. A function $f : A \to \mathsf{R}$ is *bounded* if fA has upper and lower bounds. We write $\sup_{a\in A} f(a)$ or $\sup_a f(a)$ for $\sup fA$; similarly for $\inf$.

14. An infinite set A is *countable* if there is a bijection $A \to \mathsf{N}$. The set Q of rational numbers is countable, but R is not. See e.g. Willard [150].

References

[1] Abraham, R. *Foundations of Mechanics*, Benjamin, New York 1967.

[2] Abraham, R. and Robbin, J.W. *Transversal Mappings and Flows*, Benjamin, New York 1967.

[3] Adams, J.F. Vector fields on spheres, *Ann. Math.* 75 (1962), 603-632.

[4] Adams, J.F. *Lectures on Lie Groups*, Benjamin, New York 1969.

[5] Agoston, M. *Algebraic Topology: A First Course*, Marcel Dekker, New York 1976.

[6] Ahlfors, L.V. and Sario, L. *Riemann Surfaces*, Princeton University Press 1960.

[7] Andronov, A.A. and Pontrjagin, L.S. Systèmes grossiers, *Dokl. Akad. Nauk.* 14 (1937), 247-251.

[8] Anosov, D.V. Roughness of geodesic flows on closed Riemannian manifolds of negative curvature, *Soviet Math. Dokl.* 3 (1962), 1068-1070.

[9] Arnol'd, V.I. Singularities of smooth mappings, *Russian Math. Surveys* 23 (1968), 1-43.

[10] Arnol'd, V.I. Lectures on bifurcations in versal families, *Russian Math. Surveys* 27 (1972), 54-124.

[11] Arnol'd, V.I. *Ordinary Differential Equations*, M.I.T. Press, Cambridge, Mass. 1973.

[12] Arnol'd, V.I. Critical points of smooth functions and their normal forms, *Russian Math. Surveys* 30 (1975).

[13] Auslander, L. and MacKenzie, R.E. *Geometry of Manifolds*, Academic Press, New York 1964.

[14] Berry, M.V. Waves and Thom's theorem, *Adv. Phys.* 25 (1976), 1-26.

[15] Bishop, R.L. and Crittenden, R.J. *Geometry of Manifolds*, Academic Press, New York 1964.

[16] Blackett, D.W. *Elementary Topology*, Academic Press, New York 1967.

[17] Borel, A. et al. *Seminar on Transformation Groups*, Annals of Mathematics Studies 46, Princeton University Press 1960.

[18] Bourbaki, N. *Eléments de Mathématique* 33: *Variétés Différentiables et Analytiques*; *Fascicule de Résultats*, Hermann, Paris 1967.

[19] Bowen, R. *Equilibrium States and the Ergodic Theory of Anosov Diffeomorphisms*, Lecture Notes in Mathematics 470, Springer-Verlag, Berlin· Heidelberg 1975.

[20] Brickell, F. and Clark, R.S. *Differentiable Manifolds*, Van Nostrand Reinhold, London 1970.

[21] Bröcker, Th. and Lander, L.C. *Differentiable germs and catastrophes*, L.M.S. Lecture Notes Series 17, Cambridge University Press 1975.

[22] Chern, S.S. and Smale, S. (eds.) *Global Analysis: Proceedings of A.M.S. Symposium in Pure Mathematics Vol.* XIV, American Mathematical Society, Providence, R.I. 1970.

[23] Chern, S.S. and Smale, S. (eds.) As above, Vol. XV.

[24] Chevalley, C. *Theory of Lie Groups*, Princeton University Press 1946.

[25] Chillingworth, D.R.J. Smooth manifolds and maps, in *Global Analysis and its Applications, Vol.*I. IAEA, Vienna 1974.

[26] Chillingworth, D.R.J. Elementary catastrophe theory, *Bull. Inst. Math. Appl.* 11 (1975), 155-159.

[27] Chillingworth, D.R.J. The catastrophe of a buckling beam, in [71].

[28] Chillingworth, D.R.J. and Furness, P.M.D Reversals of the earth's magnetic field, in [71].

[29] Chow, S.N., Hale, J.K. and Mallet-Paret, J. Applications of generic bifurcation, I, *Arch. Rat. Mech. Anal.* 59 (1975), 159-188.

[30] Coddington, E. and Levinson, N. *Theory of Ordinary Differential Equations*, McGraw-Hill, New York 1955.

[31] Cullen, C. *Matrices and Linear Transformations*, Addison-Wesley, Reading, Mass. 1966.

[32] Dieudonné, J. *Foundations of Modern Analysis*, Academic Press, New York 1960.

[33] Dodson, C.T.J. and Poston, T. *Tensor Geometry*, Pitman, London 1976.

[34] Dodson, M.M. Darwin's law of natural selection and Thom's theory of catastrophes, *Math. Biosciences* 28 (1976), 243-274.

[35] Duistermaat, J.J. Oscillatory integrals, Lagrange immersions and unfoldings of singularities, *Comm. Pure Appl. Math.* 27 (1974), 207-281.

[36] Eells, J. *Singularities of Smooth Maps*, Nelson, London 1968.

[37] Field, M.J. Equivariant dynamical systems, *Bull. A.M.S.* 76 (1970), 1314-1318.

[38] Franks, J.M. Ω-stability: diffeomorphisms and flows, in *Proceedings of Colloquium on Smooth Dynamical Systems* (ed. D.R.J. Chillingworth), Southampton University 1972.

[39] Franks, J. Absolutely structurally stable diffeomorphisms, *Proc. A.M.S.* 37 (1973), 293-296.

[40] Franks, J. Time dependent stable diffeomorphisms, *Inv. Math.* 24 (1974), 163-172.

[41] Golubitsky, M. An introduction to catastrophe theory and its applications *(to appear)*.

[42] Golubitsky, M. and Guillemin, V. *Stable Mappings and Their Singularities*, Graduate Texts in Mathematics 14, Springer-Verlag, New York 1973.

[43] Griffiths, H.B. and Hilton, P.J. *A Comprehensive Textbook of Classical Mathematics*, Van Nostrand Reinhold, London 1970.

[44] Grobman, D. Homeomorphisms of systems of differential equations, *Dokl. Akad. Nauk.* 128 (1959), 880-881.

[45] Gromoll, D. and Meyer, W. — On differentiable functions with isolated critical points, *Topology* 8 (1969), 361-369.

[46] Guckenheimer, J. — Absolutely Ω-stable diffeomorphisms, *Topology* 11 (1972), 195-197.

[47] Guckenheimer, J. — Catastrophes and partial differential equations, *Ann. Inst. Fourier (Grenoble)* 23 (1973), 31-59.

[48] Guillemin, V. and Pollack, A. — *Differential Topology*, Prentice-Hall, New Jersey 1974.

[49] Halmos, P. — *Finite-Dimensional Vector Spaces*, Van Nostrand, Princeton, N.J. 1958.

[50] Hartman, P. — On the local linearization of differential equations, *Proc. A.M.S.* 14 (1963), 568-573.

[51] Hartman, P. — *Ordinary Differential Equations*, Wiley, New York 1964.

[52] Hirsch, M.W. — *Differential Topology*, Graduate Texts in Mathematics, Springer-Verlag, New York 1976.

[53] Hirsch, M.W. and Pugh, C.C. — Stable manifolds and hyperbolic sets, in [22].

[54] Hirsch, M.W., Palis, J., Pugh, C.C. and Shub, M. — Neighbourhoods of hyperbolic sets, *Inv. Math.* 9 (1970), 121-134.

[55] Hirsch, M.W. and Smale, S. — *Differential Equations, Dynamical Systems, and Linear Algebra*, Academic Press, New York 1974.

[56] Hochschild, G. — *The Structure of Lie Groups*, Holden-Day, San Fransisco 1955.

[57] Hocking, J.G. and Young, G.S. — *Topology*, Addison-Wesley, Reading, Mass. 1961.

[58] Hoffman, K. — *Analysis in Euclidean Space*, Prentice-Hall, New Jersey 1975.

[59] Hurewicz, W. — *Lectures on Ordinary Differential Equations*, M.I.T. Press, Cambridge, Mass. 1958.

[60] Iooss, G. — Bifurcation of a periodic solution of the Navier-Stokes equations into an invariant torus, *Arch. Rat. Mech. Anal.* 58 (1975), 57-76.

[61] Irwin, M.C. On the stable manifold theorem, *Bull. L.M.S.* 2 (1970), 196-198.

[62] Irwin, M.C. and Robertson, S.A.R. (To appear).

[63] Keller, J.B. and Antman, S. *Bifurcation Theory and Nonlinear Eigenvalue Problems*, Benjamin, New York 1969.

[64] Kervaire, M. A manifold which does not admit any differentiable structure, *Comm. Math. Helv.* 34 (1960), 257-270.

[65] Kupka, I. Contribution à la théorie des champs génériques, *Contrib. Diff. Eqs.* 2 (1963), 457-484; 3 (1964), 411-420.

[66] Lang, S. *Introduction to Differentiable Manifolds*, Wiley (Interscience), New York 1962.

[67] Lang, S. *Analysis* I, Addison-Wesley, Reading, Mass. 1968.

[68] Lang, S. *Differential Manifolds*, Addison-Wesley, Reading, Mass. 1972. (Revised and expanded version of [66].)

[69] Lefschetz, S. *Differential Equations: Geometric Theory*, Wiley (Interscience), New York 1957.

[70] Loomis, L.H. and Sternberg, S. *Advanced Calculus*, Addison-Wesley, Reading, Mass. 1968.

[71] Manning, A.K. (ed.) *Dynamical Systems – Warwick* 1974, Lecture Notes in Mathematics 468, Springer-Verlag, Berlin·Heidelberg 1975.

[72] Markov, A.A. The unsolvability of the problem of homeomorphy, *Proc. Internat. Congress of Mathematicians* 1958, Cambridge University Press 1960, pp. 300-306 (in Russian).

[73] Markus, L. *Lectures in Differentiable Dynamics*, C.B.M.S. Regional Conference Series in Mathematics 3, American Mathematical Society, Providence, R.I. 1971.

[74] Markus, L. and Meyer, K.R. Solenoids in generic Hamiltonian dynamics (*to appear*).

[75] Marsden, J., Ebin, D., and Fischer, A. Diffeomorphism groups, hydrodynamics and relativity, *Proceedings of the 13th Biennial Seminar of the Canadian Mathematical Congress* (ed. J.R. Vanstone), C.M.C. 1972.

[76] Marsden, J.E. and McCracken, M. *The Hopf bifurcation and its Applications*, Applied Math. Series 19, Springer-Verlag, Berlin•Heidelberg • New York 1976.

[77] Massey, W.S. *Algebraic Topology: An Introduction*, Harcourt, Brace and World, New York 1967.

[78] Mather, J. Stability of C^∞ mappings:
I. *Ann. Math.* 87 (1968), 89-104
II. *Ann. Math.* 89 (1969), 254-291
III. *Publ. Math. I.H.E.S.* No.35 (1968),127-156
IV. *Publ. Math. I.H.E.S.* No.37 (1969),223-248
V. *Adv. in Math.* 4 (1970), 301-336
VI. in *Proceedings of Liverpool Singularities Symposium* I (ed. C.T.C. Wall), Lecture Notes in Mathematics 192, Springer-Verlag, Berlin•Heidelberg 1971.

[79] Maunder, C.R.F. *Algebraic Topology*, Van Nostrand, New York 1970.

[80] Mendelson, B. *Introduction to Topology*, Allyn & Bacon, Boston, Mass. 1962.

[81] Milnor, J.W. On manifolds homeomorphic to the 7-sphere, *Ann. Math.* 64 (1956), 399-405.

[82] Milnor, J.W. *Morse Theory*, Annals of Mathematics Studies 51, Princeton University Press 1963.

[83] Milnor, J.W. *Topology from the Differentiable Viewpoint*, University Press of Virginia 1965.

[84] Mirsky, L. *An Introduction to Linear Algebra*, Oxford University Press 1955.

[85] Moser, J. *Stable and Random Motions in Dynamical Systems: With Special Emphasis on Celestial Mechanics*, Annals of Mathematics Studies 77, Princeton University Press 1973.

[86] Munkres, J.R. *Elementary Differential Topology*, Annals of Mathematics Studies 54, Princeton University Press 1961 (revised 1966).

[87] Narasimham, R. *Analysis on Real and Complex Manifolds*, Masson, Paris 1968.

[88] Newhouse, S. On simple arcs between structurally stable flows, in [71].

[89] Newhouse, S. and Palis, J. — Bifurcations of Morse-Smale dynamical systems, in [97].

[90] Nitecki, Z. — *Differentiable Dynamics*, M.I.T. Press, Cambridge, Mass. 1971.

[91] Nomizu, K. — *Fundamentals of Linear Algebra*, McGraw-Hill, New York 1966.

[92] Palais, R.S. — Morse theory on Hilbert manifolds, *Topology* 2 (1963), 299-340.

[93] Palais, R.S. — Critical point theory and the minimax principle, in [23].

[94] Patterson, E.M. — *Topology*, Oliver and Boyd, Edinburgh 1956.

[95] Peixoto, M.M. — Structural stability on two-dimensional manifolds, *Topology* 1 (1962), 101-120.

[96] Peixoto, M.M. — On an approximation theorem of Kupka and Smale, *J. Diff. Eqns.* 3 (1967), 214-227.

[97] Peixoto, M.M. (ed) — *Dynamical Systems:* Proceedings of Salvador Symposium 1971, Academic Press, New York 1973.

[98] Peixoto, M.M. — On the classification of flows on 2-manifolds, in [97].

[99] Pitts, C.G.C. — *Introduction to Metric Spaces*, Oliver and Boyd, Edinburgh 1972.

[100] Poénaru, V. — *Singularités* C^∞ *en présence de symétrie*, Lecture Notes in Mathematics 510, Springer-Verlag, Berlin·Heidelberg 1976.

[101] Poston, T. and Stewart, I. — *Taylor Expansions and Catastrophes*, Research Notes in Mathematics 7, Pitman, London 1976.

[102] Poston, T., Stewart, I.N. and Woodcock, A.E.R. — *The Geometry of the Higher Catastrophes*, Research Notes in Mathematics, Pitman, London (to appear).

[103] Pugh, C.C. — The closing lemma, *Amer. J. Math.* 89 (1967), 956-1009.

[104] Robbin, J.W. — On the existence theorem for differential equations, *Proc. A.M.S.* 19 (1968), 1005-1006.

[105] Robbin, J.W. — A structural stability theorem, *Ann. Math.* 94 (1971), 447-493.

[106] Robinson, R.C. Generic properties of conservative systems, I, II, *Amer. J. Math.* 92 (1970), 562-603 and 897-906.

[107] Robinson, R.C. Structural stability of C^1 flows, in [71].

[108] Robinson, R.C. Structural stability of C^1 diffeomorphisms, to appear in *J. Diff. Eqns.*

[109] Rudin, W. *Principles of Mathematical Analysis*, 2nd ed., McGraw-Hill, New York 1964.

[110] Ruelle, D. and Takens, F. On the nature of turbulence, *Comm. Math. Phys.* 20 (1971), 167-192.

[111] Sard, A. The measure of the critical values of differentiable maps, *Bull. A.M.S.* 48 (1942), 883-890.

[112] Sattinger, D.H. (ed.) *Topics in Stability and Bifurcation Theory*, Lecture Notes in Mathematics 309, Springer-Verlag, Berlin•Heidelberg 1973.

[113] Sewell, M.J. Some mechanical examples of catastrophe theory, *Bull. Inst. Math. Appl.* 12 (1976), 163-172.

[114] Shields, P.C. *Linear Algebra*, Addison-Wesley, Reading, Mass. 1964.

[115] Shub, M. Stability and genericity for diffeomorphisms, in [97].

[116] Shub, M. Structurally stable systems are dense, *Bull. A.M.S.* 78 (1972), 817-818.

[117] Siersma, D. Classification and deformation of singularities; thesis, Amsterdam 1974.

[118] Simmons, G.F. *Introduction to Topology and Modern Analysis*, McGraw-Hill, New York 1963.

[119] Smale, S. Morse inequalities for a dynamical system, *Bull. A.M.S.* 66 (1960), 43-49.

[120] Smale, S. On gradient dynamical systems, *Ann. Math.* 74 (1961), 199-206.

[121] Smale, S. Generalized Poincaré's conjecture in dimensions greater than four, *Ann. Math.* 74 (1961), 391-406.

[122] Smale, S. Stable manifolds for differential equations and diffeomorphisms, *Ann. Scuola Norm. Sup. Pisa* 18 (1963), 97-116.

[123] Smale, S. An infinite dimensional version of Sard's theorem, *Amer. J. Math.* 87 (1965), 861-866.

[124] Smale, S. Structurally stable systems are not dense, *Amer. J. Math.* 88 (1966), 491-496.

[125] Smale, S. Differentiable dynamical systems, *Bull. A.M.S.* 73 (1967), 747-817.

[126] Sotomayor, J. Generic one-parameter families of vector fields, *Publ. Math. I.H.E.S.* No.43 (1974), 5-46.

[127] Sotomayor, J. Generic bifurcations of dynamical systems, in [97].

[128] Spivak, M. *Calculus on Manifolds*, Benjamin, New York 1965.

[129] Spivak, M. *A Comprehensive Introduction to Differential Geometry*, Publish or Perish, Boston, Mass. 1970.

[130] Stamm, E. Introduction to differential topology, *Proceedings of the 13th Biennial Seminar of the Canadian Mathematical Congress* (ed. J.R. Vanstone), C.M.C. 1972.

[131] Sternberg, S. On the structure of local homeomorphisms of euclidean n-space II, *Amer. J. Math.* 80 (1958), 623-631.

[132] Stewart, I.N. The seven elementary catastrophes, *New Scientist* 68 (1975), 447-454.

[133] Sussmann, H.J. Catatastrophe theory, *Synthèse* 31 (1975), 229-270.

[134] Sutherland, W.A. *Introduction to Metric and Topological Spaces*, Oxford University Press 1975.

[135] Takens, F. Hamiltonian systems: Generic properties of closed orbits and local perturbations, *Math. Ann.* 188 (1970), 304-312.

[136] Takens, F. Singularities of vector fields, *Publ. Math. I.H.E.S. No.* 43 (1973), 47-100.

[137] Takens, F. Tolerance stability, in [71].

[138] Thom, R. *Stabilité Structurelle et Morphogénèse*, Benjamin Advanced Book Program, Reading, Mass. 1972. English translation by D.H. Fowler, Benjamin A.B.P. 1975.

[139] Thom, R. A global dynamical scheme for vertebrate embryology, (A.A.A.S., 1971: Some Math. Questions in Biology VI), *A.M.S. Lectures on Mathematics in the Life Sciences* 5 (1973), 3-45.

[140] Thom, R. and Zeeman, E.C. Catastrophe theory: its present state and future perspectives, in [71].

[141] Thompson, J.M.T. and Hunt, G.W. *A General Theory of Elastic Stability*, Wiley, London 1973.

[142] Thompson, J.M.T. and Hunt, G.W. Towards a unified bifurcation theory, *Z.A.M.P.* 26 (1975), 581-604.

[143] Thompson, M. The geometry of confidence: an analysis of the Enga *te* and Hagen *moka*, a complex system of pig-giving in the New Guinea Highlands, to appear in *Rubbish Theory*, Paladin.

[144] Tromba, A.J. The Morse lemma on arbitrary Banach spaces, *Bull. A.M.S.* 79 (1973), 85-86.

[145] Wallace, A.H. *Differential Topology: First Steps*, Benjamin, New York 1968.

[146] Wassermann, G. *Stability of Unfoldings*, Lecture Notes in Mathematics 393, Springer-Verlag, Berlin• Heidelberg 1974.

[147] Whitney, H. Differentiable manifolds, *Ann. Math.* 37 (1936), 645-680.

[148] Whitney, H. The self-intersections of a smooth n-manifold in 2n-space, *Ann. Math.* 45 (1944), 220-246.

[149] Whitney, H. On singularities of mappings of Euclidean spaces I, Mappings of the plane into the plane, *Ann. Math.* 62 (1955), 374-410.

[150] Willard, S. *General Topology*, Addison-Wesley, Reading, Mass. 1970.

[151] Williams, R.F. Expanding attractors, *Publ. Math. I.H.E.S. No.* 43 (1974), 169-204.

[152] Zeeman, E.C. C^0 density of stable diffeomorphisms and flows, in *Proceedings of Colloquium on Smooth Dynamical Systems* (ed. D.R.J. Chillingworth), Southampton University 1972.

[153] Zeeman, E.C. On the unstable behaviour of stock exchanges, *J. Math. Economics* 1 (1974), 39-49.

[154] Zeeman, E.C. Primary and secondary waves in developmental biology, (A.A.A.S., 1974: Some Mathematical Questions in Biology VIII), *A.M.S. Lectures on Mathematics in the Life Sciences* 7 (1974), 69-161.

[155] Zeeman, E.C. Levels of structure in catastrophe theory, in *Proceedings of the International Congress of Mathematicians, Vancouver* 1974, Canadian Mathematical Congress 1975, II, 533-546.

[156] Zeeman, E.C. Catastrophe theory, *Scientific American* 234 (1976), 65-83.

[157] Zeeman, E.C. Euler buckling, in *Catastrophe Theory - Seattle* 1975, Lecture Notes in Mathematics 525, Springer-Verlag, Berlin·Heidelberg 1976.

[158] Zeeman, E.C. The umbilic bracelet and the double cusp catastrophe, in *Catastrophe Theory - Seattle* 1975, as above.

[159] Zeeman, E.C. and Trotman, D.J.A. The classification of elementary catastrophes of codimension ≤ 5, in *Catastrophe Theory - Seattle* 1975, as above.

Addenda

[160] Field, M.J. *Differential Calculus and its Applications*, Van Nostrand Reinhold, London 1976.

[161] Meyer, K.R. Generic bifurcations in Hamiltonian systems, in [71].

Index

186

298 9 1703